EARTH AND WORLD

KELLY OLIVER

EARTH AND WORLD

*Philosophy After
the Apollo Missions*

COLUMBIA UNIVERSITY PRESS

NEW YORK

COLUMBIA UNIVERSITY PRESS
Publishers Since 1893
New York   Chichester, West Sussex
cup.columbia.edu

Copyright © 2015 Columbia University Press
All rights reserved

Library of Congress Cataloging-in-Publication Data
Oliver, Kelly, 1958–
Earth and world : philosophy after the Apollo missions / Kelly Oliver
Includes bibliographical references and index.
ISBN 978-0-231-17086-4 (cloth) — ISBN 978-0-231-17087-1 (pbk.)
— ISBN 978-0-231-53906-7 (e-book)
1. Philosophy—21st century. 2. Earth (Planet)—Miscellanea. 3. Philosophy, Modern. I. Title.
B805 O46 2015
190—dc23
2014035071

COVER DESIGN: Martin Hinze

For Chuck and Teri

...thanks for our many adventures,
enjoying the wonders of earth...

From the viewpoint of the processes in the universe and in nature, and their statistically overwhelming probabilities, the coming into being of the earth out of cosmic processes, the formation of organic life out of inorganic processes, the evolution of man, finally, out of processes of organic life are all "infinite improbabilities," they are "miracles."
—Hannah Arendt, *Between Past and Future*

CONTENTS

Acknowledgments xi

1. The Big Picture: Philosophy After the Apollo Missions 1
2. The Earth's Inhospitable Hospitality: Kant 45
3. Plurality as the Law of the Earth: Arendt 71
4. The Earth's Refusal: Heidegger 111
5. The World Is Not Enough: Derrida 163
6. Terraphilia: Earth Ethics 207

Notes 247

Bibliography 263

Index 293

ACKNOWLEDGMENTS

Thanks to Jennifer Fay for ongoing discussions of Earth and World and for her work on the inhospitable world in film. Thanks to Cynthia Willett for reading the entire manuscript and giving me helpful feedback and for her inspiring work on *Interspecies Ethics*. Thanks to the graduate students in my seminar on Earth and World at Vanderbilt. And thanks to my research assistants over the years that I have been working on this project, Juliana Lewis, Melinda Hall, and Geoffrey Adelsberg. As always, I appreciate the support and companionship of my beloved Beni, Juracán, and Yukiyú, who take the edge off the solitary pursuit of writing.

EARTH AND WORLD

FIGURE 1.1. *Blue Marble.* Image by Reto Stöckli, rendered by Robert Simmon. Based on data from the MODIS Science Team

1

THE BIG PICTURE

Philosophy After the Apollo Missions

> Men's conception of themselves and of each other has always depended on their notion of the earth.
> —ARCHIBALD MACLEISH, "RIDERS ON EARTH TOGETHER, BROTHERS IN ETERNAL COLD"

The spectacular images from the 1968 and 1972 Apollo missions to the moon, *Earthrise* and *Blue Marble,* are the most disseminated photographs in history.[1] Indeed, *Blue Marble,* the most requested photograph from NASA, is the last photograph of the planet taken from outside Earth's atmosphere.[2]

Whereas *Earthrise* shows the earth rising over the moon, with elliptical fragments of each (the moon is in the foreground, a stark contrast from the blue and white earth in the background), the later image, *Blue Marble,* is the first photograph of the "whole" earth, round with intense blues and swirling white clouds so textured and rich that it conjures the three-dimensional sphere.

Even more than previous photographs of Earth, the high definition of *Blue Marble* and the quality of the photograph make it spellbinding. Set against the pitch-black darkness of space that surrounds it, the earth takes up almost the entire frame. Unlike in *Earthrise,* in *Blue Marble* the earth does not look tiny or partial, but whole and grand. Both of these photos from Apollo missions (8 and 17) were immediately met by surprise, along

2 THE BIG PICTURE

FIGURE 1.2. *Earthrise.* Image Credit: NASA

with excited exclamations about the silent beauty of this "blue marble," this "pale blue dot," this "island earth."³

In the frozen depths of the cold war, and over a decade after the Soviets launched the first satellite to orbit Earth, Sputnik, these images were framed by rhetoric about the "unity of mankind" floating together on a "lonely" planet. At the same time as vowing to win the space race with the Soviet Union, the United States wrapped the Apollo missions in transnational discourse of representing all of "mankind." Indeed, these now iconic images ignited an array of seemingly contradictory reactions. Seeing Earth from space generated new discussions of the fragile planet, lonely and unique, in need of protection. These tendencies gave birth to the environmental movement. At the same time, the Apollo missions spawned movements to unite the planet through technology. Heralded as man's greatest triumph, the moon missions lead to a flurry of speculation on not just the technological mastery of the world, or of the planet, but also of the universe. While seeing Earth from space caused some to wax poetic about Earth as our only home, it led others to imagine life off-world on other planets. While aimed at the moon, these missions brought the Earth into focus as never before.

The photographs of Earth from space sparked movements aimed at "conquering" our home planet just as we had now "conquered" space. Indeed, critics of these early ventures into space asked why we were concentrating so many resources on the moon when we had plenty of problems here on Earth, not the least of which was the threat of nuclear war (*Time* 1969). The Apollo missions were a direct outgrowth of this threat in terms not only of the significance of the race to space but also rocket technologies, which originated with military developments in World War II. The atom bombs dropped in Japan in 1945 heralded the nuclear age with the threat of total annihilation. And the development of rockets by both the United States and Germany as part of military strategies in WWII gave rise to rockets launched into space by the USSR and USA in decades that followed. Indeed, the U.S. recruited German scientists to work with NASA.

Within a decade, we had gone from world war and the threat of genocide of an entire people, to the possibility of nuclear war and the threat of annihilation of the entire human race. And, within another decade or two (with Sputnik and then the Lunar Orbiter and Apollo missions and photographs of Earth from space), the *world* gave way to the *planetary* and the *global*. Following Heidegger, we might call this "the globalization of the world picture."[4] Indeed, upon seeing the Earth from space, Astronaut Frank Borman exclaimed, "Oh my God! Look at that *picture* over there! Here's the Earth coming up. Wow, that is pretty! You got a colour film, Jim?" And William Anders responded, "Hand me that roll of colour quick."[5] The Earth had become a *picture* even before the photo was taken. Within a few short decades, the rocket science used by the military in WWII had given rise to the globalism that we have inherited today. From global telecommunications such as cell phones and Internet to global environmental movements, the Apollo missions moved us from thinking about a world at war to thinking about the unification of the entire globe. If, after seeing *Earthrise,* as Archibald MacLeish claims, "Men's conception of themselves and of each other has always depended on their notion of the earth," then, in order to understand ourselves and our interactions with other earthlings we need to ponder the meaning of *earth* and *globe* and the place of human and nonhuman worlds in relation to them.

FROM GLOBALIZATION TO EARTH ETHICS

In this book, we explore some philosophical reflections on *earth* and *world*, particularly insofar as they relate to the *globe*, and thereby to *globalization*, to begin to develop an ethics of earth, or earth ethics. Starting with Immanuel Kant, we follow a path of thinking our relations to each other through our relation to the earth. We move from Kant's politics based on the fact that we share the limited surface of the earth, through Hannah Arendt's and Martin Heidegger's warnings that by leaving the surface of the earth we endanger not only politics but also our very being as human beings, to Jacques Derrida's last meditations on the singular world of each human being. This trajectory leads us from Kant's universal laws, which apply to every human being equally because we share the surface of the earth, through Arendt's insistence on a plurality of worlds constituted through relationships between people and cultures, and Heidegger's thoroughly relational account of world and earth as native ground, to Derrida's radical claim that each singular human being—perhaps each singular living being—constitutes not only *a* world but also *the* world. In all of these thinkers we find a resistance to world citizenship and to globalization, even as some of them embrace different forms of cosmopolitanism. And yet, in their work we also find the resources to think the earth against the global in the hope of returning ethics to the earth. The very meaning of *earth* and *world* hang in the balance. So too does our relationship to earth, world, and to other earthlings.

The guiding question that motivates this book is: How can we share the earth with those with whom we do not even share a world? The answer to this question is crucial in terms of figuring out whether there is any chance for cosmopolitan peace through, rather than against, both cultural diversity and the biodiversity of the planet. Can we imagine an ethics and politics of the earth that is not totalizing and homogenizing? Can we develop earth ethics and politics that embrace otherness and difference rather than co-opting them to take advantage of a global market? How can we avoid the dangers of globalization while continuing to value cosmopolitanism? Following the trajectory from Kant's universalism to Derrida's radical singularity, we explore different conceptions of earth and world in order to develop an ethics grounded on the earth. Inspired by Kant's suggestion that political right is founded on the limited surface of the earth, and Derrida's suggestion that each singular being is not only *a* world but also *the* world, we can articulate an ethics of sharing the earth even when we do not share a world. This earth ethics is based on our shared cohabitation of our earthly home. Building on insights from Kant, Arendt, Heidegger, and Derrida, we conclude with an ethics of earth that is formed through tensions between politics as the sphere of universals and ethics as the sphere of the singular, tensions between world and earth, or what Heidegger calls "strife" between the two. Our ambivalent place on earth is signaled in that tension. The tension between politics and ethics, world and earth, resonates with the contradictory reactions to the first images of earth from space, urges to master along with humility, urges to flee along with the urgent desire for home. Acknowledging our ambivalence toward earth and world may be the first step in accepting that, in spite of the fact that we cannot predict or control either of them, we are responsible for both.

What we learn from Kant is that public right demands a universal principle of hospitality based on the limited surface of the earth. Although Kant limits hospitality to the right of visitation, still the "original common possession" of the earth's surface becomes the basis for all political rights. Taken to its logical limits, Kant's suggestion that political right is grounded on the earth as common possession undermines private property. In addition, the clear limit that Kant delineates between politics and morals begins to blur. What happens, then, when we start with Kant's insights into the connection between politics and the earth and extend them to ethics? Here, Kant's remarks on hospitality become the basis for an ethics of the inhospitable hospitality of the earth. Derrida famously extends Kant's hospitality to its limit in unconditional hospitality to the point that we can no longer defend individualism or nationalism—indeed, to the point where we can no longer distinguish between host and guest, hospitality and hostility.[6] Arendt too extends Kant's notion of cosmopolitan hospitality to ground the always shifting "right to have rights," which takes us beyond nationalism and toward our shared coexistence and cohabitation of both earth and world.[7] Although Kant is clear that cosmopolitanism is a public or political right and not a moral one, he also insists on harmony between morals and politics.[8] Contemporary criticisms of Enlightenment universalism, however, have reopened the split between morals and politics. Or, perhaps more to the point, both morals and politics are seen as at odds with ethics. If morality and political rights are necessarily universal, how can they account for singularity and the uniqueness of each individual? Derrida, among others, has tried to reformulate Kant's principles of hospitality and cosmopolitanism in order to take into consideration the singularity of each person, even of each living being.[9] The question, then, is: Can we negotiate between moral universalism and ethical particularism? Following Kant, Arendt, and Derrida, we must grapple with this tension between universal and particular, which sometimes shows up as a tension between nationalism and cosmopolitanism.

In this vein, Arendt struggles with the tension between national rights and human rights. While she embraces the plurality and unpredictability of the human condition, she insists on the necessity and priority of state rights over generic human rights. She argues that the notion of *human rights* reduces the human being to its species, like an animal, and does not have the force of any law behind it. Unlike rights guaranteed by sovereign states, human rights come with no such guarantee. While she allows for international laws that might provide such protection for stateless peoples, she also argues in favor of state citizenship and reminds us of the importance of membership in nation states, at least in a world where international law does not provide adequate protections for stateless peoples.[10] The Apollo missions with their America First rhetoric of winning the cold war along side their transnational rhetoric of goodwill to all mankind are emblematic of this tension between nationalism and cosmopolitanism.

What we learn from Arendt is that politics is not primarily a matter of law, nor of hospitality per se, but rather it is based on the plurality of our cohabitation of both earth and world, a plurality through which we can and must create equality in the face of nearly infinite diversity. Arendt's distinction between earth and world provides important resources for rethinking Kant's notions of public and private right, the tension between morals and politics, and the notions of common possession and private property, especially in relation to war. For Kant, war is the result of the fact that we share the limited surface of the earth, and yet this same fact can and should eventually lead to perpetual peace. For Arendt, the plurality and diversity of our coexistence on earth leads to war; yet for the sake of politics, and ultimately for the sake of the endurance of human worlds, war must be limited. Arendt imagines a plurality of worlds formed through human interaction and relationships. While in some sense these worlds are of our choosing, in another sense they are radically unchosen. We are born into a world, worlds, and the world. Moreover, as she tells Eichmann, none of us has the right to choose with whom to share the surface of the earth. The unchosen nature of our earthly cohabitation gives rise not only to political obligations but also to ethical ones.

And yet, as we learn from Heidegger, the givenness of our inheritance must be continually taken up as an issue for the sake of the future of each individual and each people. While it is true that we are born into a world that we do not choose and that the unchosen character of our earthly cohabitation brings with it plurality, diversity, unpredictability, and promise, in order to become properly thought it must be interpreted and reinterpreted. Going beyond Heidegger, we could say that the unchosen nature of our earthly habitation and cohabitation must be taken up in order to become properly political and ethical. In other words, the givenness of inhabitation and cohabitation is not enough in itself to ensure that we act ethically or extend political rights toward other earthlings or the earth. Going beyond merely sharing the earth's limited surface or the plurality of cohabitation, Heidegger adds profound relationality that makes every inhabitant fundamentally constituted through its relation to its environment. This interrelationality is not just the result of earthly ecosystems that sustain the life of each individual and each species. Rather, for Heidegger, this interrelationality operates at the ontological level whereby the very being of each individual and each species comes into being through relationships and relationality. With Heidegger, we get a sense of what it means to cohabit the earth as a dynamic of buzzing, flowing, raging, blowing, thanking, thinking, dwelling, and warring. If for Kant and Arendt our relations to the earth and to the world ground political right and politics itself, for Heidegger our relations to earth and world are not only constitutive of our very being but also of the ethos in which we live, all of us together in strife and wonder, cohabitants of earth. Following Heidegger, we can begin to imagine sharing the earth, even if we do not share a world. But, moving beyond Heidegger, we can begin to articulate an ethics of earth based on this sharing and not sharing.

What we learn from Derrida's last seminar is that we both do and do not share the world. Emphasizing the profound singularity of each being—not just human beings—Derrida claims that with each one's death the world is destroyed. This radical claim evokes a sense of urgency in relation to ethics and politics. Reformulating the tension between ethics and politics supposedly resolved by Kant, Derrida embraces the impossible intersection of ethics and politics, of the demand to respond to the singularity of each and the demand for justice for all. Derrida takes us beyond Kant's notion of universal conditional hospitality to singular unconditional hospitality in the face of the absence of the world, particularly the world of moral calculation. This incalculable obligation to others—even the ones whom we may not recognize—challenges all political pluralities and notions of "peoples" in whose names "we" act. Derrida's analysis brings to the fore the tension between ethical obligations that take us beyond calculation or universal principles and political obligations that necessitate calculation and universal principles. Emphasizing the radical singularity of each living being, in his last seminar Derrida employs the figure of an island. The island connotes isolation, even loneliness, cut off from civilization. Each singular life is an island cut off from all others. But even Robinson Crusoe on his desert island is not alone. The desert or deserted island is an illusion, a fantasy of autonomy and self-sufficiency. The ethical obligation that stems from the singularity of each living being is not Robinson Crusoe's solipsistic delusion of mastery or control over himself or the plants, animals, and, ultimately, other humans that both share and do not share that island. On the one hand, Derrida insists that we do not share the world and that each singular being is a world unto itself, not just *a* world, but *the* world. On the other hand, and at the same time, we are radically dependent on others for our sense of ourselves as autonomous and self-sufficient, illusions that come to us through worldly apparatuses. We both do and do not share the world. With Derrida we get a sense of ethical urgency that we must return from the world to the earth. Even when the world is gone, the earth remains. Even if we do not share a world, we do share the earth. Moving from Kant's political right founded on the limited surface of the earth, through Arendt's plurality of worlds, and Heidegger's profoundly relational account of earth and world, to Derrida's ethics that begins when the world is gone, these chapters return earth to world and thereby ground ethics on the earth.

This book brings the urgency and responsibility of ethical obligation together with a sense of home or habitat as belonging to a community conceived as ecosystem and ultimately as belonging to the earth. By emphasizing the radical relationality of each living being, together with our shared but singular bond to the earth, the island no longer appears deserted or isolated, but rather inhabited by immense biodiversity and surrounded by oceans teaming with life that we risk discounting at our peril.[11] Indeed, the island is the meeting of land and sea, an in-between space, which provokes uncanny ambivalence. When we disavow the uncanny strangeness and our own ambivalence toward it, we then risk reducing the island to the fantasy of barren isolation or exotic paradise, neither of which is sustainable. The danger of the island metaphor is that it becomes a figure for isolation rather than a figure for the in-between, the meeting of land and sea. Even Derrida's invocation of the island as a figure for singularity risks overshadowing the necessary interconnections and interrelationality through which, and in which, habitation happens. The island metaphor when extended to the earth itself discounts the very features that make the earth unique, namely the dynamic relationships between land and sea, globe and atmosphere, planet and solar system. Moving from Kant to Derrida, can we begin to articulate an ethics and politics grounded on the earth, not as isolated island or totalizing globe, but rather as the unknown and unpredictable source of diverse life on our uncanny earthly home. Earth ethics is grounded on the earth as the home to cultural diversity and biodiversity and a plurality of worlds.

WORLD WAR, EARTH ANNIHILATION

Obviously, Kant never saw the Earth from space. But that didn't stop him from imagining how it might look to alien space travelers. Kant's appeals to extraterrestrials to make various points throughout his work are infamous.[12] And, his philosophy of public right is based on thinking of the earth as a whole. Hannah Arendt was fifty-one years old when Sputnik sent photographs back to Earth, and she died three years after Apollo 17 sent back the now famous *Blue Marble* photograph. Martin Heidegger was sixty-eight when Sputnik circled the Earth, and he died one year after Arendt in 1976. News of Sputnik and photographs of Earth from space profoundly affected both Arendt and Heidegger. And both warned of the dangers of severing our connection to Earth in favor of interplanetary thinking. Jacques Derrida, who died in 2004, was only twenty-seven when Sputnik made its appearance. He gave his last seminars in Paris in 2002 just as the United States went after Al Qaeda in Afghanistan and waged war against Saddam Hussein for a second time, using the most technologically advanced military arsenal on the planet. One of Derrida's last texts published before his death was *Philosophy in a Time of Terror*, his response to the attacks on the

World Trade Center on September 11, 2001, in which he criticizes globalization for widening the gap between the rich and the poor while embracing the rhetoric of unification and equality (2003a).

While Kant could not have imagined the threats of nuclear apocalypse or global terrorism, he did argue that nation-states must not do anything in war to preclude future peace (Kant 1996c [1795]). Even while he maintained that through war human beings find political equilibrium and develop as peoples and nations, he also imagined the possibility of perpetual peace by limiting war. For Kant, war was necessary within limits. And the surface of the earth provides the absolute and ultimate limit to both individual and national expansion. In the face of the apparent limitless threat of nuclear war, however, the so-called superpowers looked to space to escape the limitations imposed by the surface of the earth. While imperialist expansion on our own planet was limited by war, the technologies of war and the race for military dominance of the world gave rise to the possibility of colonizing space. If in the eighteenth century Kant imagined the cosmopolitan unification of the world through the necessary evil of war, after the images of Earth from space, twentieth century thinkers hoped for the unification of the globe through the dystrophic fantasy of the annihilation of the entire planet. American imperialism took on tones of unifying all humankind through space exploration, even while promoting America first and dominance over the Russians and then the Chinese. On the one hand, the possibility of nuclear destruction lead us into space during the cold war and, on the other, the military technologies produced during World War II propelled us into space.

The real nuclear destruction in World War II and the threat of nuclear war during the cold war sparked fantasies of nuclear devastation in the popular imaginary. Films such as Roger Corman's *The Day the World Ended* (1956), Stanley Kramer's *On the Beach* (1959), Stanley Kubrick's *Dr. Strangelove* (1964), and the James Bond film *You Only Live Twice* (1967), along with lesser known films such as *The Final War* (1960) and *The Day the Earth Caught Fire* (1961), all revolved around the threat of nuclear destruction, many of them imagining what would happen if the USA or USSR "pushed the button."[13] The mushroom cloud became the iconic image of nuclear destruction. After the Apollo photographs of "whole" earth, the mushroom cloud and fear of nuclear destruction was joined in our cultural imaginary by images of other types of annihilation of the entire planet through our own pollution and climate change or at the "hands" of aliens, themes that continue to be popular in Hollywood film today. Some in the environmental movement spurred on by the Apollo photographs imagined the Earth itself as our enemy, taking its revenge on us by trying to eradicate us from its surface (see Lazier 2011, 619). The mushroom cloud and the iconic Blue Marble became intertwined in popular investment in the fantasy of *whole* earth.[14]

It was as if we could think the whole earth only by imaging its destruction, that all attempts to "save" the planet require first imagining destroying it. This same tendency shows up in contemporary philosophical attempts to think earth and world. For Kant, the limited surface of the earth leads to both war and peace. Arendt thinks both earth and world in relation to the nuclear threat and space-exploring technologies that threaten their destruction. As in Kant's discussion of earth, war plays a central role in Arendt's discussions of both earth and world. The genocide of World War II prompts her discussion of what it means to have a world and whether we occupy different worlds; furthermore, she considers what it means when war is aimed at obliterating an entire world, as in the world of the Jewish people. Just as Kant argued that all war must allow for peace, Arendt argues that all war must allow for the existence of a plurality of worlds. She concludes in her analysis of the Eichmann trial, no one has the right to choose with whom he or she inhabits the earth. Unlike world war as it was waged before Hiroshima, nuclear war threatens not just the world of a particular people but also the Earth itself: we have the technology to destroy the entire planet.

War surrounds Heidegger's thought in a more menacing way inasmuch as he was not only deeply affected by World War II but also joined the Nazi party. War is also central to Heidegger's thoughts on both earth and world. Taking up the Greek notion of *polemos*, however, as he does so many times, Heidegger complicates and reconceptualizes war as the strife, or constitutive tension, between revealing and concealing that ensures that both earth and world exceed technological enframing and calculation. While Heidegger warns that the technological worldview threatens earth and world annihilation, he also deploys an alternative conception of war as *polemos* that takes us beyond warring nation states toward a more primordial force of the earth itself. In his last seminar Derrida's reflections on world also lead him to war, particularly since the U.S. invaded Iraq at the same time as the final meetings of the seminar. Even before that point, however, Derrida's remarks revolve around a line from a poem by Paul Celan, "the world is gone, I must carry you," that guide his meditations on world, finitude, and solitude, ethics, death, and loneliness—a phrase that evokes the annihilation of the whole world.

It is not just in popular culture, then, that we think of the world only by imagining its destruction. In order to take the world as a whole, we imagine it gone. In order to see the whole earth, we fantasize its obliteration. In this regard, fantasies of Whole Earth and One World are nostalgic in that they begin with imaginary scenarios of annihilation followed by the longing for wholeness. In this retroactive temporality we embrace the earth and the world by first imagining them gone and then reconstructing them whole. In the words of the tagline of the 2013 film *Oblivion*, in which aliens have rendered the earth a barren desert and convinced the few remaining inhabitants, whom they have cloned, that habitable earth was irradiated through our own nuclear war, "Earth is a memory worth fighting for." Is it a stretch to say that before the world wars, we had no sense of the world as a whole? And is it just coincidence that the images of the "whole" earth appear only through the threat of nuclear annihilation of the entire planet? Are the mushroom cloud and Blue Marble two sides of the same coin, namely the technological mediation of our relationship to both warth and world? (cf. Lazier 2011, 619). Did the threat of nuclear destruction change the *world* into a *globe*? Finally, does imagining the earth as a whole necessarily mean imagining its demolition? If so, this is certainly an example of what Derrida calls the *logic of autoimmunity*, namely that what is supposed to save us, the image of the whole earth, at the same time signals its self-destruction. Or, as Heidegger might suggest, do these images of destruction also contain the "saving power"? The seeming *wholeness* of Whole Earth and One World *seen* in the photos from space is a phantasm created by the fallout of the fantasies of world being gone and earth being obliterated. Might the fear of the destruction of earth and world inspire us to attend to our ambivalent relationship to our home planet? In part, it was this fear that inspired men to reach for the stars, the fear that life as we know it on earth might disappear one day. On "the day the world ended" and "the day the earth caught fire," these men wanted to be ready to abandon ship and make a new start someplace else in the universe. Yet what they discovered, with their first ventures off world rocketing to the moon, is that looking back and seeing the Earth was the most profound moment of their mission. Certainly, the most enduring legacy of the Apollo missions remains the images of Earth from space.

"ESCAPE FROM THE PLANET THAT WAS NO LONGER THE WORLD . . ."

Immediately after the *Earthrise* photograph was transmitted back to Earth from Apollo 8 on Christmas day 1968, poet Archibald MacLeish's wrote an article in the *New York Times* entitled "Riders on Earth Together, Brothers in Eternal Cold" in which he claimed the significance of the moon mission as changing our very conception of earth: "The medieval notion of the earth put man at the center of everything. The nuclear notion of the earth put him nowhere—beyond the range of reason even—lost in absurdity and war. . . . To see the earth *as it truly is*, small and blue and beautiful in that eternal silence where it floats, is to see ourselves as riders on the earth together, brothers on that bright loveliness in the eternal cold—brothers who know now they are truly brothers" (MacLeish 1968, my emphasis). MacLeish speculates that seeing the earth "as it truly is" will "remake our image of mankind" such that "man may at last become himself" (1968). Seeing the earth "whole" for the first time unites all of "mankind," together on "that little, lonely, floating planet." Realizing that we are all in this together on the precarious lovely earth alone in the "enormous empty night" of space is seen as a catalyst for our finally coming into our own as a species united as "brothers." When MacLeish calls the astronauts "heroic voyagers who were also men," however, we cannot hear the universal mankind, let alone humankind, but rather the masculine heroic space cowboys who have the power and vision to unite all men as "brothers" against the eternal cold of space.[15]

MacLeish's assessment is consistent with NASA's press releases after both Apollo missions, which included panhuman themes of uniting mankind and representing all of mankind in space outside of any national borders.[16] For example, then NASA chief Thomas Paine told *Look* magazine that photographs of Earth from space "emphasize the unity of the Earth and the artificialities of political boundaries" (quoted in Poole 2008, 134). NASA presented the Apollo 8 mission as one of peace and goodwill to all mankind (cf. Cosgrove 1994, 282). And, in 1969 *Time* named Apollo 8 astronauts Borman, Lovell, and Anders Men of the Year. The accompanying article described a new world born from their mission, one in which the human race could come together with one unified peaceful purpose as a result of the "escape from the planet that was no longer the world" (*Time* 1969). The world had expanded to include the universe, while the earth had shrunk into a tiny fragile ball.

Time describes the earth as a troubled place full of war and strife and space as the great hope to "escape the troubled planet." Again, the astronauts are seen as heroic figures conquering space: "It seemed a cruel paradox of the times that man could conquer alien space but could not master his native planet" (1969). The goal is clearly to conquer; and the Apollo missions signal a great victory in escaping a warring planet and moving beyond what appear from space as the petty disagreements between peoples. In the words of astronaut Frank Borman, "when you're finally up at the moon looking back at the Earth, all those differences and nationalistic traits are pretty well going to blend and you're going to get a concept that this is really one world and why the hell can't we learn to live together like decent people" (quoted in Poole 2008, 133–134). The irony is that Borman claims that he only accepted the mission because as a military officer he wanted to "win" the cold war (see Poole 2008, 17). The *Time* article concludes that man will not turn into "a passive contemplative being" because he knows how to challenge nature; and in reaching for the moon he now conquered not only the seas, the air, and natural obstacles, but also space and the moon, which brings with it the "hope and promise of his latest conquest" (*Time* 1969, 17). Like Borman, the American media seemed to think of the Apollo mission as a triumph for freedom and hope, paradoxically both for "all of mankind" and as an American victory in the cold war (cf. Poole 2008, 134). The Apollo missions, then, are emblematic of our ambivalent relationship to the earth and to other earthlings. The reactions to seeing the Earth from space make manifest tensions between nationalism and cosmopolitanism and between humanism, in the sense that we are the center of the universe, and posthumanism, in the sense that we are insignificant in the universe. In these reactions to seeing the Earth, there are contradictory urges to both love it and leave it.

"ONE WORLD" AND "WHOLE EARTH"

Ideals of One World and Whole Earth, born out of the Apollo missions, make manifest some of these tensions that signal our ambivalent relation to the earth. The *Earthrise* and *Blue Marble* photographs became emblems of both ideals. One World is the idea that technoscience can unite all of the nations of the world, while Whole Earth is the idea that concern for the shared environment can unite all peoples on the fragile planet Earth. Whereas for One World advocates the photographs of Earth from the Apollo missions signify "secular mastery of the world through spatial control," for Whole Earth advocates they represent "a quasi-spiritual interconnectedness and the vulnerability of terrestrial life" and "the necessity of planetary stewardship" (Cosgrove 1994, 287).[17] Whereas One World is the image of the entire planet connected through technology, Whole Earth is the image of the entire planet interconnected organically through the uniqueness of Earth's fragile biosphere.

The sentiments of Buckminster Fuller epitomize the One World view of the Earth as a technological wonder that unites mankind: "our space-vehicle Earth and its life-energy-giving Sun and tide-pumping Moon can provide ample sustenance and power for all humanity's needs" (quoted in Lazier 2011, 284). While this utopian vision is far from realized on Spaceship Earth, the technological goal of uniting the planet through telecommunications is increasing its reach, if not across the entire globe then certainly across a sector of the earth determined by access to technology through wealth. Still, early telecommunications originated in military operations. The first satellites were cold war technologies developed to spy on the enemy. The One World reaction with its ideal of global technology comes out of the conquering model voiced by astronauts and media alike in response to the cold war. For, even more than most, the cold war was about technology and the race to the moon was a battle over the future of technoscientific progress. According to *Time*, if we could conquer alien space, then we should be able to master our native planet (*Time* 1969).

The Whole Earth reaction, on the other hand, with its notions of organic interconnectedness and the vulnerability of our native planet, comes from the "loneliness" of Earth seen floating in the "vast night" of space. Perhaps more surprising than One Worlders' reveling in the technocratic triumph of the moon missions was the solemnity of realizing that Earth is the only body that looks even remotely alive from that vantage point. While it is not so surprising that astronauts may have felt alone floating in their space capsule thousands of miles from any other living soul, it is remarkable that their sense of isolation was contagious. Each one of them expressed the loneliness of space, from which Earth appears as an oasis. For example, on a later mission, Apollo 11 astronaut Michael Collins voiced the loneliness and vulnerability of Earth when circling the dark side of the moon alone in the Command Module: "I am alone now, truly alone, and absolutely isolated from any known life. . . . Then, as the Earth rose over the lunar horizon: so small I could blot it out of the universe simply by holding up my thumb. . . . It suddenly struck me that that tiny pea, pretty and blue, was the earth . . . I didn't feel like a giant. I felt very, very small."[18] Collins's remarks express the contradictory reactions to seeing Earth from that distance. On the one hand, he imagines blotting out the Earth with his thumb and on the other he imagines himself as very small and insignificant. The power and mastery of technological prowess is counterbalanced by the vastness of the universe that makes our entire planet look like a "tiny pea."

The first astronauts to circle the moon all expressed similar sentiments, emphasizing the loneliness, uniqueness, and fragility of Earth. Apollo 8 mission commander Frank Borman called Earth "a grand oasis in the big vastness of space."[19] And astronaut James Lovell described the loneliness of space, "The vast loneliness up here is awe-inspiring, and it makes you realize just what you have back there on earth. The earth form here is a grand ovation to the big vastness of space."[20] While Astronaut William Anders stressed the fragility of the tiny planet, referring to Earth as a "fragile Christmas-tree ball which we should handle with care."[21] To these astronauts, and subsequently the media, along with One World and Whole Earth proponents, the Earth is alone in the universe, "a planet so eccentric, so exceptional" that the mission to the moon brought the Earth into focus (Lazier 2011, 623). Through the lens of the Apollo cameras, the lovely planet Earth appears as lonely as it is unique set against the absolute blackness of space. Seeing the Earth from space, so tiny, and yet the only visible color, prompted ambivalent feelings of vast loneliness and eerie insignificance along with immense awe and singular importance.

"THIS ISLAND EARTH"

> Dreaming of islands . . . is dreaming of pulling away, of being already separate.
> —GILLES DELEUZE, "DESERT ISLANDS"

Speaking of the Apollo 8 astronauts in a publication entitled *This Island Earth*, NASA echoes the cosmopolitan sentiments of the unification of mankind, stressing that from the vantage point of space we see the "true reality" of our situation: "Their eyewitness accounts impressed millions of men with the true reality of our situations: the oneness of mankind on this *island Earth*, as it *floats* eternally in the silent sea of space."[22] Resonant with the astronauts, along with writers like MacLeish, another NASA administrator hopes that "By heeding the lessons learned in the last decade, and attacking our man made problems with the same spirit, determination, and skill with which we have ventured into space, we can make 'this *island earth*' a better planet on which to live."[23] The comparison between the earth and an island works to highlight the supposed "reality" of our situation, all in the same boat, so to speak. Like an island, the Earth is imagined alone, floating in the infinite sea of space. And, like so many fantasy islands, some see it as paradise, while others can't wait to escape from their exile here. Seeing Earth from space made some appreciate Earth anew, while others imagined moving further away from Earth and traveling to other planets. For some, seeing the loveliness of Earth "is to wish also to return" to it (Lazier 2011, 620); while for others, seeing the insignificance of Earth compared to the vastness of space is to wish to leave it. And, more to the point, the ambivalence between loving and leaving it shows up in various discourses around the first images of earth from space. In other words, these seemingly contradictory impulses, triggered within the cultural imaginary by the Apollo photographs, make manifest a deep ambivalence in our relationship to our earthly home. We feel both marooned and miraculous.

Indeed, in almost Heideggerian terms, mission chief Frank Borman recounts feeling nostalgic and homesick when he saw the "picture" of Earth from the moon: "It was the most beautiful, heart-catching sight of my life, one that sent a torrent of nostalgia, of sheer homesickness, surging through me."[24] Certainly, the photographs of Earth from the moon continue to provoke the uncanniness of our relation to our home planet, particularly when we realize that we are "down there" somewhere, minuscule specks on that tiny "pale blue dot" floating in space. Carl Sagan's recitation from the introduction to his *Pale Blue Dot* sends shivers up the spine, as he reminds us that the tiny dot, barely visible from the vantage point of space, is our home and home to everyone who has ever lived. Just watch a YouTube video of Sagan's presentation of *Pale Blue Dot* to see what I mean. Sagan concludes, "There is perhaps no better demonstration of the folly of human conceits than this distant image of our tiny world. To me, it underscores our responsibility to deal more kindly with one another, and to preserve and cherish the pale blue dot, the only home we've ever known" (Sagan 1994, 7). Sagan's book embraces the ambivalence of both love it and leave it, as the singularity of our home planet shines through the darkness of space and at the same time that vantage point reminds us of the adventurousness of the human spirit.

Even while encouraging us to love our home planet, Sagan would no doubt embrace Buzz Aldrin's recent call to colonize Mars and become a "two-planet species." "Our earth," says Aldrin, "isn't the only world for us anymore. It's time to seek out new frontiers" (Aldrin 2013). While some, like Lovell, see the Earth from space and want to protect it, others, like Aldrin, imagine escaping from Earth to find our way in the galaxy, perhaps even in the universe. With environmental disaster looming large on the horizon, in recent years there is a sense among some that the Earth has betrayed us or is taking its revenge on us, and, rather than a safe haven, it has become a death trap and a threat to human survival (see Lazier 2011, 619). The urge to colonize Mars or find another habitable home is getting stronger, evidenced by the Mars One project, which plans to start colonizing Mars in 2023, less than a decade from now, and to continue bringing people on a one-way trip to Mars every two years from then on, for a permanent self-sustaining Mars settlement.

Although it is Robinson Crusoe's island and not the earth that Derrida has in mind when he asks why some people love islands and are drawn to them while others fear islands and flee them, his analysis of the ambivalence conjured by islands is apt when considering contradictory reactions to seeing the earth from space. Indeed, on Derrida's analysis the island becomes a figure for this ambivalence, for loving or leaving or, more accurately, both loving *and* leaving. The figure of the island represents seemingly contradictory desires, namely to love and to leave, to cling to and escape from, to stand your ground and to flee. In this regard, the figure of the island might be what Julia Kristeva would call *abject* in the sense that it cannot be categorized.[25] Rather it is always in between, between land and sea, between safe harbor and threatening isolation. Like everything abject, islands both fascinate and terrify. They fuel our ambivalent and contradictory desires to love and leave, to stay and to flee.

Derrida describes these ambivalent desires in Robinson Crusoe's relationship to his desert island. The imagined circularity of Robinson's island comes to represent a turn inward as a return to self. Even as Robinson longs to escape his family and England, he recreates their laws and customs on his island. In this sense, both literally and metaphorically, he recreates the wheel. Like Robinson's wheel and his island, the earth spinning on its axis incites ambivalence. These contradictory reactions exist within one and the same nation, within one and the same person. And yet this ambivalence now figured by the island—this island earth—undermines the sovereignty of that nation and that person. In Derrida's terms they contain within them the logic of autoimmunity that makes all *autos*, all returns to self, eventually turn against themselves, and then what was intended to protect the self begins to destroy it. This mythical return to self contains within it a threat to self, the threat of self-annihilation. For example, technologies developed to provide endless energy deliver nuclear destruction. Or fossil fuel–driven technologies that allowed us to leave earth's atmosphere and visit the moon contribute to the destruction of earth's atmosphere and perhaps eventually the destruction of life on earth.

THE VIEW FROM SPACE

Certainly, in the case of contradictory reactions to the *Earthrise* and *Blue Marble* photographs, this ambivalent logic is operative as we acknowledge the singularity of our earthly home, and at the same time attempt to escape from it to find another home. Moreover, this autoimmune logic is intrinsic to the photographs themselves. For, in order to shoot those images, astronauts were propelled into inhospitable space in an unsustainable and precarious artificial environment where their very survival was uncertain. In other words, those images could only be taken from a vantage point where the survival of man is impossible. This extraterrestrial vista is from an impossible viewpoint where no one could live. In this way the photographs signal the danger inherent in the viewpoints of the people taking them. On the one hand, these two photographs, taken by human beings rather than unmanned satellites, have more rhetorical force because they are tokens of a human eyewitness standpoint. On the other, they also signal the perilous position of these space travelers who risk their lives while taking them. The only way to get what even NASA officials called this "God's eye view" was from an impossible point so far away from Earth.[26] The view of the "whole" Earth could not be seen from Earth, but only at a distance born out of rocket science and compared to the viewpoint of God. As creatures on the earth, we cannot see the Earth; it is never a whole or total object presented to our perception. Apart from photographs, the view of the Earth as a whole has been reserved for the rare astronaut who left the Earth's atmosphere. Speaking of their view of "the whole globe," as "the first humans to see the world in its majestic totality," astronaut Frank Borman exclaimed, "This must be what God sees," and many of the astronauts talked of traveling to "the heavens."[27] To see the Earth whole, as it "really is," human beings must travel to the heavens to get a God's eye view of the planet. Some earthlings even asked the astronauts whether they had seen God in space (Poole 2008, 129). The view of the whole Earth is the view of God. It is no wonder, then, that the Apollo missions sparked as much discussion of conquering and mastery as they did vulnerability and fragility.

What the astronauts, the media, and the One World and Whole Earth proponents assumed they saw in the photographs, particularly *Blue Marble,* namely the whole Earth, however, was an illusion. For both images show only part of Earth, indeed, a fraction of the Earth. *Earthrise* shows an elongated piece of the top of a sphere, while *Blue Marble* shows one side of the Earth; and both are rendered in the two-dimensional space of the photographic medium. In other words, we *did not* see what we thought that we saw. The impact of seeing the Earth whole, seeing it as it *really* is, was based on the fantasy of the Whole Earth, which not only was never visible in these photographs but also, at least with current technology, never will be. The Whole Earth cannot be captured from any human vantage point, even one floating in a space capsule orbiting the moon or any other point in space. For, as phenomenologists teach us, the human perspective is always only partial; there is always something that is occluded and missing from our viewpoint.[28]

DOES THE EARTH MOVE?

Decades before the first Apollo mission, Edmund Husserl wrote an essay that he placed in an envelop with the title "Overthrow of the Copernican theory in usual interpretation of a worldview. The original ark, earth, does not move" written on the outside.[29] There, Husserl maintains that in our everyday experience, the earth does not move but is rather that in relation to which everything else moves. Speaking phenomenologically, we do not directly experience the spinning of the earth on its axis as movement. Husserl begins with the observation that, for us, "the earth is a globe-shaped body, certainly not perceivable in its wholeness all at once and by one person; rather it is perceived in a primordial synthesis as a unity of mutually connected single experiences" (Husserl 1981a, 222–223). Husserl goes on to argue that in order to have a notion of absolute motion and rest, we must take the earth as the basis-body (which is indeed no body, no thing, at all) as our ground, or what he calls *the earth-basis* and *earth-basis arc*: "Although for us it is the experiential basis for all bodies in the experiential genesis of our idea of the world. This 'basis' is not experienced at first as body but becomes a basis-body at higher levels of constitution of the world by virtue of experience" (Husserl 1981a, 222–223). In our everyday experience we do not perceive the earth as a body or a thing, but rather as the ground for our perception of all other bodies or things.

Husserl argues that even if we leave the earth and travel in space, perhaps even colonize other planets, the earth is still the basis of our experience insofar as we are born here and everything we know and experience is relative to life on earth. He says, "All of that is relative to the earth-basis ark and 'earthly globe' and to us, earthly human beings, and Objectivity is related to the All of humanity" (1981a, 228). He rejects the idea that human life could have sprung up on Venus or the Moon, suggesting that if there is life there it won't be human life as we know it. We are essentially earthly creatures whose history and experience is inherently linked to the earth. The earth as the basis of that history and experience is not just one body among others, a planet like any other. Rather, it is our home. One of Husserl's main arguments is that science should not forget that the earth is pregiven for us, and therefore all our understanding, knowledge, and theories are born out of that pregivenness. Or, as he puts it, "All brutes, all living beings, all beings whatever, only have being-sense by virtue of my constitutive genesis and this has 'earthly' precedence. . . . There is only one humanity and one earth" (1981a, 230). The earth is the primordial ark that makes everything else possible for us.[30] The function of the earth for Husserl is to bear and support us not only as our original home but also as the original body against, and through, which we experience our own bodies and the world. While it is true that in one sense the photographs of earth from space show us one body among others, this misses the point that we comprehend all other bodies in space only in relation to our earthly experience. We understand all other bodies in relation to the earth as our primordial ground, our basis-body.

Furthermore, the photographs of earth only show one portion of earth, one viewpoint on earth, and never the whole earth. This is because there is no possible human viewpoint from which to see the whole earth. That vantage point is impossible for us. Thus, if by "God's eye view" we mean seeing the whole, then we can never adopt the position of God. Moreover, whatever we see has its origins in our earthly existence and is relative to the earth as the basis of our perception. Thus we can never see the earth "as it truly is" unless what we mean by "as it truly is" is what it is for us. Even then, post-Husserlian phenomenology has challenged the notion that there is universal meaning for all human beings. If we can only ever take a perspective on something, then it is possible that we may have different perspectives. Insofar as the earth is not a something like any other *thing*, but rather the ground for our perception of every thing as a something, then the problem of perspective becomes even more complex. If the earth is the basis for any and all of our perspectives, and even for our ability to take perspectives, then it is pregiven in a way that, as Heidegger says, is always concealed from us. Additionally, the pregivenness of the earth may be different for different peoples or cultures, which Heidegger suggests is linked to traditions and ultimately the earth as the "native ground" for "a people." This is not to say that *native ground* or *a people* are homogeneous or universal either. For, as Derrida drives home, the phenomenological insight contains within it the autoimmune logic such that if we have perspectives all the way down, so to speak, then the constitutive genesis of the ego as transcendental, or universal, along with its native ground, begins to tremble. Our world, if not the very earth itself, begins to quake.

Both One World and Whole Earth impulses were based on the illusion that from space we could get a God's eye view on the Earth as a whole. In this regard, in spite of their differences, they share the assumption that the photos show the totality of Earth as it "truly is." The two are totalizing and global ideologies that promote managing the earth or mastering it, for the sake of global technology in the case of One World and for the sake of the global environment in the case of Whole Earth. Geographer Denis Cosgrove argues that both these totalizing movements depended on seeing the Earth whole, or more to the point, imagining seeing the Earth whole: "*Seeing* the Earth whole is critical to the imaginative reception of the space images and to the totalizing socio-environmental discourses of One-world and Whole-earth to which they have become so closely attached" (Cosgrove 1994, 271).

But, we have not seen the Earth whole. And we have not seen the Earth *as it truly* is, isolated from everything around it. Indeed, without its atmosphere, the Earth would not look like the beautiful blue marble of the photographs. Furthermore, the Earth looks beautiful and unique relative to the black space around it, the gray surface of the moon, and the reflection of light from the sun. This is to say, the photographs are not just images of the Earth alone, but also the Earth in relation to the elements that surround it and constitute the Earth as more than just a planetary body. To take the Earth as an object apart from its relationships is the ultimate illusion of the mastery of our own subjectivity, a subject so powerful and grand that it can take the whole Earth as it object. As Heidegger might remind us, the Earth, even as seen from space, only appears in relationship to other elements, whatever those elements may be. In Heidegger's terms, those elements include the sky (perhaps atmosphere), mortals (the finite limited human perspective of the astronaut's holding the cameras), and divinities (perhaps what even the astronauts refer to as *the heavens*). To see Earth as an object floating alone in nothingness is to interpret the photographs within the technological framework that renders everything, even Earth itself, as an object *for us*, an object that can be grasp, managed, and controlled, an object ripped from its contextual home.

We haven't seen the Earth whole and (unless something drastically changes in the constitution of human perception) we never will. As phenomenology teaches us, we never see any object in its entirely; rather, ours is always and only a perspectival seeing. If I cannot see what is closest to me, my own body, whole, then how can I expect to see the entire planet Earth whole? If we think that perhaps proximity is precisely the problem and, in order to see the whole, we need to get more distance and, like the astronauts floating in space, move further away from our object, we should think again. For, while changing proximity for distance can give us a different perspective, which may be enlightening, it does not give us a view of the whole. Whether we are looking at a table and chairs a few feet away or the Earth from space, we see only one side, one perspective, and cannot, and never will, see the whole in its entirety, *as it truly is*.

Thus we must question our investments in the whole earth fantasy. The wholeness supposedly seen in the first photographs of Earth from space is reminiscent of what psychoanalyst Jacques Lacan calls the *misrecognition* of wholeness in the mirror stage.[31] Lacan describes the infant recognizing himself in the mirror for the first time and the ensuing fantasy of wholeness through which it compensates for the fragmentation of its experience. The image in the mirror presents a stable picture that the infant latches onto in order to prop up a sense of itself as autonomous and whole. And yet this is just a fantasy inasmuch as the infant's experience is far from being either autonomous or whole. Indeed, as Lacan describes the scene, the infant is held there by its mother, or another caregiver, because it is still too young to stand in front of the mirror on its own. The mirror stage becomes emblematic of this imaginary process through which we attempt, never successfully, to compensate for our own fragmentation and lack of wholeness. The fantasy or misrecognition of wholeness provided by our imaginary projection onto the image in the mirror operates as a defense against the fact that our experience is always fragmented. We are never completely autonomous or whole, which causes anxiety, which in turn, we ward off with fantasies of wholeness. Visual images are the perfect medium to allow for this sense of stability, especially photographs insofar as they are frozen in time and space. And, perhaps more than any other medium, photographs present us with the illusion of stable presence outside the vagaries of time and space.[32]

Psychoanalysis might interpret the fantasy of wholeness upon seeing the photographs of earth from space as a defensive reaction against the sense of fragmentation that we actually experience.[33] Given the turbulence of the late 1960s and early 1970s in the U.S., the Apollo photographs, along with the fantasies of the "unity of mankind," One World, and Whole Earth that they fuel, act to quell anxieties about the possibility of nuclear war and civil unrest. On the other hand, the environmental movement, hatched in the wake of these photographs, signals an investment in saving the earth from the devastation caused by humans. Certainly, many are invested in "saving the planet" by correcting or compensating for environmental damage humans have caused that may destroy our atmosphere, dramatically change our climate, and perhaps even render the earth uninhabitable for us. The fantasy of the whole earth allows us to continue with illusions of mastery and globalism, but now in the service of saving, rather than destroying, the planet.

Still, the notion that we can control and master our environment and our globe are part and parcel of the technological worldview that got us into an environmental mess in the first place. The mastering gaze that imagines itself taking the whole Earth as its object, through the Apollo photographs, perpetuates and emboldens a notion of human subjectivity as standing apart from its objects—in this case the earth—and over against them as the subjects controlling the destiny of those objects. Imagining the Earth as a body amongst others, as an object of our perception like others, is to imagine seizing it, controlling, and making it our own. But, as the most rudimentary foray into phenomenology reveals, we never see the whole of any object, but rather arrive at our sense of wholeness through processes of induction and deduction that are in themselves born out of our embodied experience as earthlings.

In the context of the cold war, and in the face of the threat of nuclear and environmental disaster, the photographs of Earth from space fuel a defensive reaction against such dangers through this fantasy of wholeness. Perhaps this is why *Blue Marble* is the most reproduced photograph in history, and why the image of the whole Earth has become the symbol for globalization.[34] The danger of globalism and planetary thinking, as Kant, Arendt, Heidegger, and others point out, is the homogenization of the world into a globe and, more perilous, into a "world picture."[35] A globe connotes something that we can control and manage, like the manufactured globes that we can hold in our hands, globes that we produce and possess. The photographs of earth from space turn earth into a globe that we imagine we can control and possess. Global thinking emerges after these first images of Earth from space. And with it comes totalizing discourses of uniting the entire planet through technology or through environmental management. Subsequently, global economy and global markets attempt to unite all humankind through consumerism, which not only incorporates and assimilates differences but also makes everything fungible. Everything has a price tag. Even saving Earth is reduced to offsetting one's "carbon footprint" by paying for it. This totalizing sense of globalization shrinks earth's cultural diversity and biodiversity as global technologies and economies expand. As Heidegger warned, the technological imperialism of the planet turns the earth into a globe, and we no longer dwell on our earthly home. Rather, globalism threatens the *desertification* of both earth and world. The totalizing of this global technological worldview operates as if it had no limits and without remainder. And this is the danger, that the plurality of the earth and of the world(s) will be reduced, assimilated, and ultimately annihilated.

THE DANGERS OF GLOBALIZATION

Hannah Arendt criticizes privileging this view of earth from space, the "Archimedean" vantage point from which we can supposedly see our world and ourselves as they *truly* are. She argues that the Archimedean point in space will necessarily keep shifting, ever more distant from earth, as part of the fantasy that if we get enough distance on our home planet then we will see it as it *really is* and then we will be able to unlock its secrets. Once we find a point from which to view the earth, we will need to move onto another one from which to view that point and the whole, ad infinitum until the "only true Archimedean point would be the absolute void behind the universe" (Arendt 1954, 278). Arendt is critical of "our modern longing to escape what some call our imprisonment on earth" (Arendt 1958a, 1). From the distance of space, human activities, she warns, will not be recognizable as such; rather, we will look like rats, and our cars will look like snail shells or turtle shells attached to our backs (Arendt 1958a, 323, 1954, 278). On her view, if our behavior is compared to that of other animals' behavior then we have been degraded. Of course, this assumes a problematic hierarchy of species.

Arendt contends that we are earthbound creatures who mistakenly see ourselves as dwellers of the universe (Arendt 1958a, 3). We imagine that we could live in that Archimedean point, off world, apart from earth. We labor under the illusion that because a few astronauts visited space we can live there. We believe that we can occupy the position from which to get "God's eye" on our home planet. Arendt suggests that this race to escape the earth is an attempt to escape the human condition, which she claims is essentially terrestrial. Echoing Husserl, she maintains that our thinking is earthbound no matter where we are in space: "the human brain which supposedly does our thinking is as terrestrial, earthbound, as any other part of the human body. It was precisely by abstracting from these terrestrial conditions, by appealing to a power of imagination and abstraction that would, as it were, lift the human mind out of the gravitational field of the earth and look down upon it from some point in the universe, that modern science reached its most glorious and, at the same time, most baffling achievements" (Arendt 1954, 271). Her choice of the word *baffling* is interesting in that it means both *bewildering* and *deceptive*, even connoting *cheating*. To call the greatest achievements of science, particularly in terms of the "conquest of space," baffling, is to suggest not only that they are bewildering or confusing but also that they are deceptive or frustrating. Perhaps even more baffling for Arendt is that the general reaction to these achievements is not bafflement, but rather matter-of-fact acceptance. The "step back," as Heidegger calls it, that allows us to pause and think about ourselves and about our relation to earth and space requires a certain bafflement or bewilderment in the face of technological advances. Without it, we risk becoming complacent with a technological worldview that is so compellingly deceptive it tricks us into thinking that we have arrived at the objective perspective—what Arendt calls Einstein's "free floating observer in space"—that Archimedean point, from which we see the earth and the world as "it truly is."

Arendt is deeply critical of what she sees as an the impulse to create an artificial world to replace the natural one, in large part because she views it as an attempt to master the earth and the world by claiming the power to create even a second moon, and perhaps even a second earth and a second world. Resonant with Heidegger's criticisms of technology, Arendt claims that the desire to escape the earth and create a new one someplace else is not only the result of denying the human condition at our own peril but also evidence of a dangerous hubris. The illusion that we can master the earth, our world, and space, leads to unchecked development and deployment of dangerous technologies that threaten all life on earth. Arendt sees this hubris in fantasies of space colonization and creation self-sustaining atmospheres for us elsewhere. Commenting on Sputnik, Arendt wrote to her mentor Karl Jaspers: "Most honored friend—What do you think of our two new moons? And what would the moon likely think? If I were the moon, I would take offense."[36] And the very first words of *The Human Condition* refer to Sputnik as the most important, and the most dangerous, event in human history.

We might wonder what Arendt would make of Biosphere 2, the self-enclosed artificial environment set in the Arizona desert where in 1991 eight people were locked in to see if they could survive for two years independently (they couldn't), or current projects like the Netherlands based Mars One, which plans to colonize Mars by 2023 with funding from a global reality TV show documenting the astronaut selection process, for astronauts willing to make the one-way journey to the red planet. Will the Mars One project fulfill Buzz Aldrin's dream of making human beings a two-planet species? Or, like Biosphere 2, will it show that human beings have only one habitable home, Biosphere 1, the Earth? Whatever happens, Husserl and Arendt would presumably share the belief that, at least until human beings are born and raised on Mars, they will still be earthlings, measuring everything according to Earth standards and using their terrestrial brains and bodies to understand their Martian environment. Furthermore, they will still have originated on Earth, which gave them life and sustained them so that they could explore space. And if they can make a life on Mars, and even if they find other life on Mars, insofar as they have a history and a past, a given from which they cannot escape, they will still be *of earth*.

Given that for Arendt Sputnik shares the same desire to control the natural world as totalitarianism, it is likely that she would be appalled at both Biosphere 2 and Mars One.[37] Perhaps even more threatening than the replacement of natural with artificial, Arendt warns against the globalism inherent in both Sputnik and totalitarianism. Both, she suggests, manifest a desire to take over the entire globe, the same desire evident in the One World and Whole Earth movements. Arendt is more than skeptical about such movements, which claim to unite the entire world or the whole earth. Although Arendt thinks that international law is necessary to protect against genocide, she rejects appeals to abstract human rights or any type of world government (see Arendt 1992, 273). She argues that the rights of man, or human rights, are too abstract to protect individuals; rather than protect them, rendering persons *merely human* reduces them to their species, which does not guarantee political rights (Arendt 1966, 300). For Arendt, each individual needs the protections of state citizenship, even if international law is also needed to protect entire states from each other.

Kant too warns against one world government, even as he argues for perpetual peace as a result of cosmopolitanism. This is why he proposes a federation of states that would maintain their own governments and also be governed by international law. For both Arendt and Kant, totalitarianism is a danger of globalism. In various ways both argue that we need the checks and balances of the possibility of war to make sure that any one state does not take over the world. Heidegger's forceful delineation of the dangers of technological globalism is well known. On his account, globalism is inherent in the technological way of framing the world with its desire for world domination. Even Derrida, who extends Kant's conditional hospitality to unconditional hospitality, rejects globalism.[38] He rejects even the words *globalism* and *globalization*. Like his predecessors, Derrida warns of the totalizing nature of globalism, which seeks control of the whole world without remainder.

This tendency reaches a certain zenith with terrorism insofar as terrorism is not tied to nation-states or citizenship, but is rather global. Terrorism is global in the sense of being able to strike almost anywhere and in the sense of threatening to destroy the entire world. In one of his last published works before his death, Derrida argues that now, with seemingly global access to technologies of destruction, the nuclear threat can come from anywhere. Unlike state violence, the terrorist threatens to strike almost anywhere at any time and threatens "the very possibility of world order." Derrida claims that what is at stake is nothing less than the world itself, the worldwide or global insofar as all of life and the very planet are at risk.[39] The globalization of technologies of destruction threatens the entire world, without remainder. Just as Arendt worries about the totalitarian tendencies of globalism, and Heidegger warns of the totalizing discourse of technology as globalism, Derrida suggests that terrorism is an outgrowth of the logic of autoimmunity operating within technoscience that makes the destruction of the whole world possible. But now, in addition to the threat of nuclear war between nation-states, we have the threat of a terrorist atomic bomb or nuclear attack that, thanks to the proliferation of technology, could strike at any time and in almost any place on earth.

Technoscience has changed our relationship to both Earth and World. Now, thanks to technology, we not only can see the planet from space, but also can destroy it. Technology changes the relationship between the earth and the world. Once man ventured into space, the earth was no longer the world. Now that man has the power to destroy the earth, the illusion that man is the master of both earth and world threatens the destruction of both. Certainly, how we conceive of the earth is intimately linked to how we conceive of ourselves. If we see the earth as an object or a resource for our use, then we see ourselves as subjects and masters. If we see the earth as a pawn in a war or terrorism, then we see ourselves as sovereigns, even gods. But, if we see the earth as our singular home, then we see ourselves as earthlings who are profoundly dependent upon the earth. Moreover, we realize that we share the earth with all other earthlings such that we are not their masters, nor the master of earth. Rather than think our world limitless, and revel in the expansion of our powers, we understand that we are limited creatures and that our world(s) are fundamentally limited by their necessary relation to earth. As Heidegger cautions, technology changes the relationship between earth and world such that world threatens to overshadow and assimilate all of earth. This fantasy inherent in the technological worldview, the fantasy of wholeness and totalization, threatens not only our ways of life but also all of life on earth.

Derrida too warns that technology has changed meaning of earth as terra: "The relationship between earth, terra, territory, and terror has changed, and it is necessary to know that this is because of knowledge, that is, because of technoscience. It is technoscience that blurs the distinction between war and terrorism" (Derrida 2003a, 101). Technology equalizes access to weapons of mass destruction. And yet, even as technology seems to give equal access to all, it simultaneously increases inequities between peoples and nations. This is why Derrida rejects the term *globalization*, because it implies that the entire globe has access to communication technologies or military technologies or so-called global markets (2003a, 121). But, this is far from true. As Derrida points out, "From this point of view, globalization is not taking place. It is a simulacrum, a rhetorical artifice or weapon that dissimulates a growing imbalance, a new opacity, a garrulous and hypermediatized noncommunication, a tremendous accumulation of wealth, means of production, technologies, and sophisticated military weapons, and the appropriation of all these powers by a small number of states or international corporations" (2003a, 123). At the same time that globalization appears to be equalizing access to global technologies or global markets, it is also increasing the divide between "haves" and "have nots." For example, just a few countries control over 80 percent of the wealth, which is up considerably in the last few decades.[40] And, as a result of globalized markets, the top three billionaires in the world have more assets than the combined GNPs of all the least developed countries and their six hundred million inhabitants—three people have more than six hundred million people (Steger 2003, 105). The notion that globalization is good for everyone is a myth.

Resonant with Heidegger, Derrida also insists that worldwide markets or technologies are not global insofar as *the world* is not synonymous with *the globe*. In order to maintain the distinction between *world* and *globe*, Derrida says, "I am keeping the French word *'mondialisation'* in preference to 'globalization' so as to maintain a reference to the world *monde, Welt, mundus*, which is neither the globe nor the cosmos" (2002a, 25).[41] While *globe* refers to the entire planet, *world* more narrowly connotes the world of human beings, or perhaps the worlds of particular species of beings, and maybe even the unique world of each singular living being. Indeed, *world* allows for multiple worlds constituted by cultural and historical differences (among others), whereas *globe* suggests the entire planet in its universal context, that is to say in the context of the universe. Although Derrida does not object to *globe* or *global*, as Heidegger does, because planetary thinking is at odds with the earth, he is clear that he thinks that *worldwide* is more accurate than *global*. Just as we interpreted the photographs of Earth from space as pictures of the whole earth, what we call *global* or *globalization* are merely fantasies of planetary wholeness. And, as Heidegger argues, the fantasy of planetary totality is dangerous when it positions itself as the only way to relate to the world.[42] The danger of this totalizing discourse is that it does not allow for differences, or even history, but rather insists on dominating everything that is. *Earth and World* is an attempt to offset what Heidegger calls the *planetary imperialism* of global technology and the totalizing technological worldview with earth ethics based on the singular bond to the earth that we share with all earthlings. Earth ethics operates as an antidote to both Heidegger's nightmare of planetary imperialism, on the one hand, and Derrida's fantasy of disconnected islands, on the other.

DESERT ISLAND

> An island doesn't stop being deserted simply because it is inhabited.
> —GILLES DELEUZE, "DESERT ISLANDS"

It is noteworthy that within the metaphorics of the philosophical trajectory of these thinkers, the technological impulse to cover the entire planet reduces the world to a *desert*. While Kant merely mentions the desert as an inhospitable region across which, thanks to the fortunes of providence, camels can transport us as the "ships" of the desert, the figure of the desert becomes a metaphor for the inhospitable regions of political geography for Arendt. Arendt describes the *worldlessness* of isolation and lack of connection between people and peoples as a desert. In that desert of worldlessness, she claims that, like a "sandstorm," totalitarianism can take over (Arendt 2005, 201). Without relationships between people, the world becomes meaningless, and without meaning it is not a world. Rather, without meaningful relationships we are left worldless. And the further danger is that totalitarianism will step in to fill this void of meaning. Arendt claims that philosophy, art, and love are "life giving sources that let us live in the desert without drying up." We need them to have the energy to rekindle the political (Arendt 2005, 202). But without these meaning-giving activities we live in the uninhabitable region of worldless desert. Even worse, warns Arendt, is when we take this worldless desert for the only world and call it home, that is to say when we adjust to "life" in the desert.

Arendt introduces the desert metaphor in the epilogue to reflections on contemporary war as threatening the destruction of the entire world: "who could deny that the conditions of the arms race under which we live and have to live at least suggest that the Kantian statement that nothing should happen in a war that makes a later peace impossible has likewise been set on its head, so that we live in a peace in which nothing may be left undone to make a future war still possible" (Arendt 2005, 200). It is not just a matter of making war such that we can eventually establish peaceful relations, but making war such that there is a world left to make more war in the future. But, even without total annihilation, war can turn the world into a desert by threatening to undermine the very possibility of politics (Arendt 2005, 154). Totalitarianism is one such threat insofar as it is a worldless world, one without politics in Arendt's sense as plurality and freedom. Her desert metaphor, then, is not postapocalyptic, but rather, as she says, "worldlessness that results from the two fold flight from the earth into the universe and from the world into the self" (Arendt 1958a, 6). Out of this desert, the world must be renewed and reborn through the oases of philosophy, art, and love that sustain us and give us the strength to recreate political bonds. Arendt proposes that *amor mundi*, or love of the world, can act as an antidote to the worldlessness of the desert of nihilism.

Arendt asks whether we love the world enough to take responsibility for it. We might ask whether we love the earth enough to take responsibility for it. We might ask whether in addition to *amor mundi* we also need *amor terra*, love of the earth. Not in terms of controlling, managing, or mastering earth, but rather, as Heidegger suggests, in terms of "cultivating and caring [*Pflegen und Hegen*]," which can also be translated as *tending* and *fostering*.[43] Heidegger insists that this cultivating and caring, tending and fostering, preserving and protecting are not forms of production (Heidegger 1951). Rather, they are ways of listening and responding to the earth's sheltering and nourishing powers. Acknowledging our singular dependence on the earth might be a starting point for caring for it. Disavowing our singular bond to the earth can lead to worldlessness, which both Arendt and Heidegger associate with the desert. For Arendt, the desertification of the world results when we lose the human relationships that make life meaningful. When we disavow our relations to other people the meaning of life becomes impoverished. The autonomous self-contained subject that denies its fundamental relationality may find itself cut off from the source of meaning, living as if on a desert island. Without meaningful relationships, we lack creativity, which Arendt associates with philosophy, art, and love. And without creativity and meaning, life is diminished as if in a desert. Going beyond Arendt, however, it is not just human plurality, nor our bonds to other people or peoples, that make life meaning but also the cultural diversity and biodiversity of life on earth and our bonds to other earthlings. Although Arendt insists on our singular bond to the earth, she overlooks the necessity of biodiversity when she claims that plurality is the law of the earth. Turning to Heidegger's notion of worldlessness helps return the (human) world to the earth in ways that begin to acknowledge our connection to both the earth and other earthlings.

Heidegger warns of worldlessness or the unworlding of the world. So, too, he warns of technology uprooting, or unearthing, us from the earth. This is what he fears when he comments that the images of earth from *Lunar Orbiter I* scare him because these images give us an earth that "is no longer *the earth* on which man lives" (Heidegger 1981 [1966], 105). Heidegger suggests that modern technology results in the flight from both earth and world. He too associates the unworlding of world with the desert, or the *desertification (Verwüstung)* of the world.[44] He claims that the "desolation of the earth" begins when we willfully employ technology without knowing its consequences, especially when we are sure of ourselves and confident in our scientific calculations or metaphysical representational thinking (1973a, 110).

In a posthumously published work, Heidegger stages a conversation in a prison of war camp in which his two characters discuss the desertification of the world that renders us worldless and conclude, "Devastation [or *desertification, Verwüstung*] means to us that everything, the world, the human, the earth, is transformed into a desert [*Wüste*]."[45] In a provocative essay on Heidegger and terrorism, Andrew Mitchell describes the unworld become desert as remaining a world, but an impoverished world.[46] Perhaps, we could

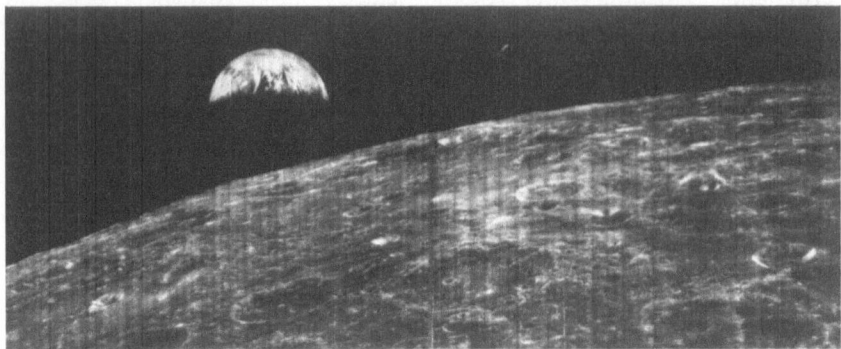

FIGURE 1.3. *Lunar Orbiter 1*. Photo Credit: NASA/LOIRP

say that this unworlding of world does not reduce us to stones, which have no world, but rather makes us more like animals, with what Heidegger calls their impoverished world. Like animals, we become poor in world when the totalizing technological worldview renders our world a meaningless desert, as if devoid of life, or at least devoid of the meaning of life. Devastation or desertification is even worse than total annihilation because at least with annihilation there is nothing left to worry about or to protect (see Mitchell 2005). Desertification, on the other hand, leaves us in an unworlded world and uprooted from the earth such that everything is standing reserve and therefore disposable. Everything, then, is destructible and therefore terrorizible. This is why Heidegger concludes, "devastation [desertification] is more uncanny than annihilation."[47] With desertification, technically we are still alive, but in a lifeless and meaningless world. Of course, the desert metaphor assumes a human perspective from which deserts appear barren and lifeless, which is far from true. The recent film *Oblivion* (2013) gives us an image of the earth that has been turned into a lifeless desert after we "used the nukes" to fight off an alien invasion. The lifeless desert of earth is still a world, but one whose meaning is precariously tied to uncanny, even illegitimate, memories of what earth once was. Within the bleak world of this film, the protagonist, Jack Harper (Tom Cruise) finds meaning in a tiny plant that has survived. This otherwise seemingly insignificant life-form becomes a symbol of hope within Harper's daily shuttle between his two worlds, the world of the ruined lifeless earth and the artificial world of his high-tech "home" suspended outside of earth's poisoned atmosphere. Within the universe of *Oblivion*, technology and the technological worldview have taken over and destroyed the earth. The alien invaders are disembodied computer intelligence with only an artificial voice that controls the environment for the only humans left. Although it seems that only two humans are left, draining the earth of its last water to render its total desertification complete, through various plot twists, we discover that even these two, including Harper, are clones engineered by the alien invader. In this film, technology not only threatens all life on earth but also threatens something even worse, namely oblivion. It is the memories of the human survivors that the aliens attempt to erase. Oblivion, or forgetfulness, is the greatest danger posed by the totalizing technological worldview.

The tagline for the film, "the earth is a memory worth fighting for," takes on a new meaning in view of the danger of forgetfulness. Harper struggles throughout the movie to remember his past, which turns out to be a past that was not really his. Through visions and flashbacks, Harper recovers his life on earth. And Harper exploits the limits of his disembodied high-tech supervisor, who still needs eyes and ears on the ground, so to speak. He flies under the radar, literally, and discovers an isolated oasis in the desert that earth has become; and there he recreates his home. One lesson that we might draw from this filmic representation of the desertification of the world that renders humanity worldless is that a return to earth and its biosphere gives our protagonist's world meaning. World is dependent on earth. And earth populates world not only with life but also and moreover with meaningful life. It is the earth that makes a world home.

A would-be Robinson Crusoe, in the end Harper escapes from his artificial island "home" in space and returns to his oasis refuge in the desert on earth, but only after encountering his own "footprint in the sand." Like the moment when Robinson discovers the footprint in the sand, the moment when Harper encounters his clone is an uncanny turning point. His life both is and is not his own. His memories both are and are not his own. Although, as a clone, Harper is replaceable and fungible, his love for his wife—whom he eventually remembers and finds alive in a stasis tube (long story)—is real and compelling. Clone or not, this love gives his life (or lives) meaning. Embracing—or perhaps just ignoring (it's Hollywood after all)—the autoimmune logic of encountering himself elsewhere, which decenters any claim that he might have to sovereignty or original agency, Harper(s) defeat(s) the technological enemy and return(s) home to the memory of earth encapsulated in the evergreen forest oasis of life and meaning in the desert the earth has become. In this science fiction fantasy, the world, and everything in it, including humans, has become fungible and destructible, everything, that is, except earth itself.

If, we live in an age of terrorism where everything is fungible and destructible, including life itself, then, following Derrida, we could also say that we live in an age of autoimmunity where life turns against itself. In his final seminar, Derrida associates the logic of autoimmunity with Robinson Crusoe's *desert island*. Just after asking his seminar audience which book they would take with them to a desert island, the Bible or a Heidegger seminar, Derrida turns to the moment in the text where Robinson Crusoe is terrorized by the sight of a human footprint in the sand. Robinson Crusoe is haunted by this footprint in part because he doesn't know if it is his own or belongs to someone else. He doesn't know if he is just retracing his steps without knowing it or if, metaphorically as well as literally, his desert island is not so deserted after all. Perhaps he is not alone. The footprint, however, is uncanny not because it may belong to someone else, but rather because it may belong to Robinson Crusoe, because "fundamentally, he cannot decide if this track is his own or not" (Derrida 2011, 48). The island, the circle, the wheel spinning on its axis become figures for the logic of autoimmunity, and, like Robinson Crusoe, we not only retrace our steps without even knowing it but we are also haunted by the other within that may yet annihilate us. In this regard, every living being is a world made up of diverse living organisms living together in one body that operates as a semicontained, but fundamentally interconnected, ecosystem or world.

With the footprint, the question whether or not Robinson Crusoe's island is actually deserted becomes the turning point in the novel. This question is also pivotal for Derrida insofar as he asks what it means to be alone and whether or not the beasts are alone. He suggests that each singular being is like an island, a world unto itself. And yet he also insists that each singular being is fundamentally dependent upon others. In this regard, whether we are talking about life with other human beings or with nonhuman animals (and perhaps even plants and rocks), we both share, and do not share, a world or *the* world.

In his earlier work the figure of the desert takes on a double meaning. For example, in "Faith and Knowledge" Derrida talks of the desert as a place where barrenness is a lack of predetermined meaning, language, or symbolic systems. Yet this barren desert before meaning is the intersubjective space that makes meaning possible. The barrenness of this desert is what enables the fecundity of life as we know it, that is to say, the possibility of meaningful life. In this desert there is no dogma, no traditions, no prejudice. There is just the primary relationality that makes communication possible. In other words, this desert is the precondition of meaning, language, and symbolic systems, which can become deserted in the Arendtian or Heideggerean sense of worldlessness or lifelessness when they tip into dogma and prejudice. What Arendt and Heidegger associate with meaninglessness when the human world is overtaken by the technological worldview Derrida associates with the conditions of possibility of meaning. What Derrida calls the "desert within the desert" refers to the performative space necessary for—and necessarily outside of—the law properly conceived, including the laws of grammar, language, and other symbolic systems or institutions, along with civil and criminal law. In this work the desert is both a hopeful place (or place before place) that allows us to presuppose the possibility of communication through a primary relatedness and a violent place where the force of law, its performative conditions, gives way to law proper. For Derrida, the desert is not only both a hopeful and a violent place but also a fecund place that gives birth to human meaning and human law.

We can relate this thinking of the desert to Derrida's analysis of Robinson Crusoe's desert island, particularly insofar as he raises the question of whether or not Robinson Crusoe is alone. Is the island really deserted? As Derrida's analysis suggests, it is only deserted from the perspective of the white male European. Obviously the island is populated with various species of animals and plants. Moreover, Robinson Crusoe eventually sees native people on the beach, whom he immediately imagines killing or enslaving. Derrida, however, goes further to show how Robinson Crusoe brings his European parents, values, and traditions with him such that his island is populated even with those he left behind. Furthermore, through his interactions with various beasts on the island, Robinson Crusoe finds himself and makes himself "sovereign." His sovereignty is the product of, and dependent on, the others with whom he cohabits his so-called desert island. Derrida complicates any notion that Robinson Crusoe is alone, even from the beginning, of his sojourn on the desert island. On his analysis, Robinson Crusoe is both alone and not alone, both singular and yet sharing a world.

And, certainly, whether or not Robinson Crusoe shares a world with the coinhabitants of his island, he does share this bit of earth, connected as it is, to the rest of our unearthly earthly home.

EARTH AS EXCESS AND LIMIT

In light of our analysis, and continuing this line of thought, we might ask, is the earth alone? And, are we alone on earth? Certainly, seeing the earth from space, and seeing photographs of earth from space, triggered thoughts of both the loneliness of earth and the uniqueness of earth. In addition, seeing the earth from that great distance provoked a sense of unity, of planetary wholeness, one world and whole earth. It spurred both fantasies of aliens seeing earth for the first time and a sense that we are truly alone in the universe. After the first moon mission, Astronaut James Lovell told *Time*, "What I keep imagining is that I am some lonely traveler from another planet. What would I think about the earth at this altitude? Whether I think it would be inhabited or not" (*Time* 1969, 16). In addition to the God's eye view, the view of Earth from space conjures the perspective of aliens "seeing" our planet for the first time.[48] What would an extraterrestrial see when "he" saw Earth? Would he think it was inhabited? Centuries before the Apollo moon missions, Kant imagined the perspective of aliens seeing the Earth. For Kant, it is from this alien vantage point that we see ourselves as earthlings united on the limited surface of the same planet. We can only see ourselves whole by imagining that we are someone, or something, else. We see ourselves as a species only when compared to other species. Perhaps this is why, without other species on earth, humans would be very lonely—that is if we could imagine the existence of humans without other species, which is impossible. And, for Kant, it is only by imagining what we look like from the perspective of another species that we might finally see ourselves as rational beings among other rational beings, citizens of the universe. We see ourselves as inhabitants of one planet only by imagining the perspective of an extraterrestrial and seeing ourselves through its "eyes." Thus, in order to "see" ourselves whole, we must split our perspective and take up what we imagine to be the perspective of another. That is to say, we fragment ourselves in order to see ourselves whole. As we have seen, reactions to photographs of earth from space manifest this same autoimmune logic. Only by taking an impossible, unsustainable, and even deadly perspective do we see ourselves as a species united on our home planet. Only from this impossible "God's eye" vantage point do we "see" ourselves whole. With these photographs from space we imagine that we have finally seen the whole earth, "as it truly is," an island floating in space. And as an island it provokes ambivalence, wanting to flee it and wanting to protect it.

The photographs of earth from space provoke the "love it or leave it" reaction that feeds the illusion of control and mastery by suggesting that we must, or can, choose one or the other, but not both. And yet, as we will continue to see, our relationship to earth is uncanny, not just because "seeing" it incites contradictory reactions but also because of our singular and pregiven bond to it that makes it always necessary both in excess of our perception and understanding and, at the same time, a limit to who we are and what we can do. We belong to the earth, and it does not belong to us or at least not in the sense of a possession and not solely. As earthlings, we depend on the network of relationships that constitute earth.

If our conception of earth determines our conception of ourselves and vice versa, then we must rethink the metaphor of earth as island. The island metaphor suggests that the earth is isolated, insular, and self-contained. The initial reaction to the photographs of earth were illusions in part because they perpetuated the view of earth as isolated, insular, and self-contained. This reaction not only reflected the tendency toward nationalist isolationism but also the individualism prominent in American culture inherited from philosophies political philosophies such as Hobbes's and Locke's. If we conceive of the earth as a network of dynamic, living, relationships, then we also change the way that we "see" ourselves. No longer maverick space cowboys desperately holding on to the fantasy of the autonomous self-contained individual, species, nation, we can embrace "plurality as the law of the earth." This means that individuals, species, and nations are fundamentally interconnected. It means that relationality is primary rather than secondary to who we are and what we can do. It means that if earth is an island, then we must reconceive islands as dynamic spaces constituted by their relationships to air, sea, and the elements that make them what they "truly are." In other words, it means that we must embrace the fact that we are limited creatures who are not just living on earth, but rather part of the biosphere that constitutes its very being. And, as such, the earth and other earthlings are always necessarily in excess of our world(s) and function as limits to our world(s). Earth ethics demands that we acknowledge the ways in which we do not share a world because the earth and other earthlings refuse us. With equal force, earth ethics demands that we acknowledge the ways in which we do share worlds because the earth and other earthlings respond to us. Ultimately, earth ethics demands that we not only acknowledge but also embrace the vital fact that, although we may not share a world, we do share a singular bond to the earth.

Earth ethics, then, acts as a counterbalance to globalism insofar as it is grounded on the earth as a dynamic network of relationships through which each and all earthlings share the earth even if they do not share a world. At bottom, the earth is a limit against which totalizing tendencies of global technology abut. The earth always and necessarily juts through the globe to remind us of our own limits. Earth ethics, then, is an ethics of limits. It is an ethics of remainder, what cannot be assimilated. Contrary to globalism, with its limitless appetite for new markets and consumption, the earth's biosphere is a delicate balance of reproduction and consumption that never completely uses itself up. What do we make of the limits of the earth and of our own limitations as earthlings bound to this our home planet? Does the globular shape of the earth lead to the Kantian insight that the surface of the earth and its resources are limited and therefore must be shared? Does it lead to the Heideggerian insight that we must limit ourselves, and our consumption, so as not to use up earth? Or, on the contrary, does it lead to war over these limited resources to the point of self-destruction, actualizing what Derrida calls the logic of autoimmunity? Does seeing the Earth limit our own desire for mastery or fuel it?

For Heidegger, the danger, along with the saving power, lies within the essence of the technological worldview. There is an autoimmune logic inherent in the planetary imperialism of global technology itself, such that it reaches its limits and turns against itself. The very danger of man-made climate change and environmental disaster may unite us not only as a species but also as earthlings with a common purpose to save our earthly home. Heidegger suggests that human beings are tasked with protecting what he calls the "self-secluding" power of the earth. And when he introduces what he calls the *fourfold*, the concealing and secluding powers permeate not just earth but also sky, mortals, and divinities. For Heidegger, these secluding powers are forces that operate beyond intentionality and synthesis and resist recuperation, incorporation, and comprehension. They are the forces that prevent us from mastering earth or world, technology or the globe. Once we realize that the totalizing tendency of the technological mode of relating to earth and world threatens to make everything, even human beings, disposable, there may be hope of finding alternative modes of relationship. Once we see that technology is but one way, and not the only way, of approaching the world, then other possibilities open up. And when we allow that there are multiple ways of relating to the world, already we begin to thwart the dangers of totalization. On the other hand, the global connections provided through technology may allow for unification with a common purpose rather than assimilation into a common market.

Earth ethics, grounded in responsibility to the earth and earthlings, expands cultural diversity and biodiversity rather than closing them down. But this opening is also the acknowledgment that human beings are not the purpose of the earth and that we do not control or master it. In this sense the globe is limited by the earth, which is beyond our control or mastery. The limits of the globe mirror our own limits, inasmuch as we realize that we always get only a partial picture and never a picture of the whole as *it truly is*. Rather, our partial and limited perspectives open up the possibility of relationships with those around us along with earth and world. The earth is constituted by a nearly infinite multitude of worlds, all partial, limited, and ultimately determined by their singular bond to the earth. Whereas the global or planetary taken as totalizing closes off relationships by fixing them into one, and only one, possible worldview, embracing our own limitations allows us to acknowledge our dependence on others, and on the social bonds, that enable belonging to both earth and world. Indeed, the planetary or the global operates in opposition to earth and world, if both are conceived as deeply relational and dependent upon cultural and biodiversity.

In sum, the guiding questions throughout this book are: Can we learn to share the earth even with those with whom we do not share a world? Can the earth operate as a limit to globalization as total assimilation?[49] Can a philosophy of the earth help us formulate an ethics of earth that does not incorporate and synthesize everything into a homogeneous whole or reduce everything to disposable resources? Following a trajectory from Kant's universal rights grounded on the limited surface of the earth, through Arendt's plurality of worlds and Heidegger's self-secluding earth, to Derridean ethics grounded on the singularity of each life as *the* world, we will explore *earth* and *world*. Building on these cumulative insights into cohabitation on the same planet, we can begin to rethink the relationship between earthly ethics and worldly politics. Following this path enables reformulating witnessing ethics in terms of addressability and response-ability now grounded in our shared, but singular, bond to the earth and the diversity of worlds that constitute our earthly home. While in earlier work I describe the strife between ethics and politics in terms of the productive tension between subjectivity and subject position, here I look to strife between earth and world (see Oliver 2001). Moving through Kant's universal public right based on hospitality and the limited surface of the earth, and Arendt's notion of plurality of cohabitation and worlds, to Heidegger's ontology of creaturely relationality and Derrida's hyperbolic ethics of responsibility to each, the ethics and politics of response-ability become intimately and inherently earthbound.

The goal of this book is to explore the concept of earth in order to ground an ethics on our singular bond to our earthly home. If there is a "saving power" of globalization, it is the potential to unite earthlings in a common purpose, attending to the health of our earthly home. Earth ethics is grounded on the responsibility to the earth and its creatures born out of the tension between the singularity of each living being, on the one hand, and our shared unchosen belonging to the same planet, on the other. This sense of belonging conjures the archaic meanings of *going along with* and *longing*, rather than belonging as *possession*. If, as Kant maintains, the limited surface of the earth is our common possession, both the notion of *common* and of *possession* must be rethought. For the earth is common not only to human beings but also to all earthlings who share a singular bond to our common home. And we belong to the earth just as much as it belongs to us, but not as a possession. Rather, this sense of earthly belonging is uncanny insofar as it necessitates the unchosen and unpredictable character of our earthly cohabitation. When we imagine ourselves masters of the earth, and we imagine other creatures as resources to be exploited, we disavow not only the singularity of each but also its singular bond to the earth, a bond that we share. Moreover, we ignore, at our own peril, the ecosystems that sustain us through cultural diversity and biodiversity. If global technologies and market-based forces threaten our earthly existence, attention to the strife between worldly politics and an earthbound ethics helps recover a sense of our earthly belonging as longing for home. Acknowledging and embracing our singular dependence on earth and its biosphere turns us from mastery over, toward care for other creatures and our shared earthly home.

Moral and political science framed as a system of knowledge, rules, or universal principles and exercised through individual or national sovereignty risks fostering the dangerous attitude of supercilious mastery and control over both earth and world. The possibility of ethics begins where this totalizing worldview ends. Earth ethics is not a system of moral rules or universal principles that we can know through reason and exercise through an autonomous will. Rather, earth ethics is responsiveness to others and to the environment by virtue of which we not only survive but also thrive. This responsiveness is based on our earthly existence as embodied creatures sharing a planet, even when we do not share a world. Rooted in the earth's unearthly strangeness, which can never be known or mastered, this ethics necessarily takes us beyond reason or recognition and toward poetic dwelling and the responsibility of love. Certainly, exploring our conception of earth tells us something about our conceptions of ourselves and about our relations to others. And traversing the relation between earth and world(s) shows us that, while earth's biosphere is made up of countless individuals, communities, societies, cultures, and worlds, we all share a singular bond to our home planet.

With hope, we embark on this voyage in order to navigate sharing the earth, even if we do not share a world. Inspired by Arendt when she says, "education is the point at which we decide whether or not we love the world enough to assume responsibility for it and by the same token save it from that ruin which, except for renewal, except for the coming of the new and young, would be inevitable," this project explores the point at which we decide whether or not we *love the earth enough to assume responsibility for it.*

2

THE EARTH'S INHOSPITABLE HOSPITALITY

Kant

In his most widely read essay, "Perpetual Peace" (1795), Kant proposes that the third definitive article of perpetual peace is based on the right to hospitality (*Hospitalitätsrecht*), which is "the right of a foreigner not to be treated with hostility because he has arrived on the land of the other" (Kant 1996c [1795], 328–329, 8:357). This hospitality or "hospitableness" (*Wirtbarkeit*) is not the right to be a guest (*Gastrecht*) because that requires entertaining the foreigner and taking him into one's own house; rather, Kant describes it as the right to visit (*Besuchrecht*) in order to present oneself to society for the sake of establishing commerce. Anyone has the right to visit any other territory to try to seek or attempt commercial relations with its inhabitants, but that is the limit of the right of hospitality (1996c, 329, 8:358). In other words, the "native inhabitants" are obligated only to listen to the proposal of the foreigner before they evict him from their territory.

What Kant means by hospitality, then, is akin to that provided by the Dutch innkeeper with whom he begins the essay; namely to accept all visitors as a business transaction.[1] Indeed, the word that he uses most frequently, which is translated as hospitality or hospitableness, is *Wirtbarkeit* (cf. *Unwirtbarkeit* or *Unwirtbarste*, translated as "inhospitable"), whose root *Wirt* means innkeeper or landlord. Thus, it has much stronger associations with the hospitality of a landlord than the English word hospitality suggests; perhaps it is helpful to think of it in terms of how the word is used today in the locution "the hospitality industry." The connections between hospitality, landlords, and property are significant not only because the right of hospitality works to guarantee the possibility of commerce between nation-states but also because the "common possession of the earth's surface" around which this right revolves is a pivotal aspect of Kant's doctrine of right or civil law, particularly property law, in *The Metaphysics of Morals* (1797; see 1966c, 329, 8:358 and 1966a, 386–511, 6:229–378).

The right to hospitality may facilitate commerce, but for Kant its source is an a priori principle of practical reason based on the limited surface of the earth. The right of hospitality is grounded on not only the spherical shape of the earth but also human dependence on the earth, which is both logically and chronologically prior to commerce as well as necessary for it. And while Kant entertains his reader with speculations about the physical, empirical, and historical chronology of relations between nations—commercial relations—his goal is to establish the metaphysical—or we might say conceptual—conditions of possibility for those relations. In *The Metaphysics of Morals* Kant bases both the right to hospitality and the right to private property (and arguably all of public right) on two facts: namely that the earth is a sphere or a globe and that all human beings live on this same planet. Thus we are compelled to ask: What does Kant means by "earth"?[2] And, what is the human being's relationship to it? Which leads us to ask, what does Kant mean by "human being"?

THE EXCESSES OF EARTH IN KANT'S PHILOSOPHY OF PROPERTY

Kant argues that the right of hospitality as (what we might call in Star Trek jargon) the right of "first contact" is necessary so that "distant parts of the world can enter peaceably into relations with one another, which can eventually become publically lawful and so finally bring the human race ever closer to a cosmopolitan constitution" (1996c, 329, 8:358). Only through a cosmopolitan constitution that unites all nation-states under one universal international law can we achieve peace, which Kant maintains, is the goal of reason itself: "establishing universal and lasting peace constitutes not merely a part of the doctrine of right but rather the entire final end of the doctrine of right within the limits of mere reason" (1996a [1797], 491, 6:355).

Why? Because "the condition of peace is alone that condition in which what is mine and what is yours [property] for a multitude of human beings is secured under *laws* living in proximity to one another, hence those who are united a constitution" (ibid.). The central problem for establishing any civil law, or the very possibility of civil law, and thereby any nation-state or international federation of states, is determining and justifying private property or distinguishing, as Kant says, what is mine and from what is yours.

One of the most curious aspects of Kant's theory of private property is his argument that both its determination and its justification lay with the finite and determinant surface of the earth. His notion of right, then, is literally grounded on the ground. That the finite surface of the earth is where humans live—it is our habitation or habitat—and that all human beings live on the earth becomes the foundation for property law, and following that all civil law, which is to say civil society or civilization itself. In other words, for Kant, in a significant sense, politics is both possible and necessary because humans are earthbound beings. Because we are earthlings, we necessarily live together on the same planet and must develop laws for sharing our common habitat, which means that we must develop laws for sharing the earth and its resources. The ultimate aim of reason is to develop universal laws that govern all human beings across the globe so that we can all share in these resources equally. And although equality is one of the three pillars of citizenship (along with freedom and independence), and by extension of cosmopolitanism—or literally "world citizenship" (*Weltbürgerrecht*)—Kant is clear that the right to equality does not mean equity; indeed, he argues that inequality among men is necessary for the development of human perfection. Given that for Kant equality does not and cannot guarantee equity or equal shares, what does he mean by our common possession of the earth? And for Kant what principles ought to guide sharing or, more to the point, dividing up this so-called original common possession? If we follow Kant's reasoning to its logical conclusion, that is to say to the ends of the earth, we discover not only that the concept of private property begins to evaporate when exposed to the elements, including the molten matter at the earth's core, but also that human hospitality is possible only because of the earth's inhospitable hospitality. The earth and its dynamic surface cannot be contained within Kant's theory of property even as it functions as the foundation for it. Moreover, the hospitality of our own bodies to the faculties of reason that distinguish us from all other animals on earth is dependent upon the earth and our cohabitation with those other earthly creatures. We are not the only earthlings. This is particularly significant insofar as Kant insists that it is not as individuals that we can or will realize the perfection of reason, but only as a species.[3] We are a species whose uniqueness is that we are "inhabited," as Kant says, by reason only in relation to other uninhabited or uninhabitable species on earth. Yet within Kant's conjectural fancies human beings become earthlings united in their possession of this planet by virtue of our place in the universe as imagined from other planets. We become earthlings only when viewed from the perspective of extraterrestrial (rational) beings. Finally, Kant's theory of cosmopolitanism taken to the ends of the earth should lead us to believe not only that peace is an ongoing, if not infinite, process that even may take us off world, but also that, both conceptually and empirically, definitive property is impossible without perpetual peace. In the end, inspired by Kant yet moving away from ownership and toward another sense of belonging, away from the distribution of resources and toward communal sharing, we might think of politics and ethics grounded on the fact that we belong to the earth, which we share with all other earthlings.

KANT'S DOCTRINE OF RIGHT AS THE DOCTRINE OF PRIVATE PROPERTY

Kant insists that every individual ought to leave the state of nature and enter into the state of civil society. This "duty" is not just the result of the fact that human beings need to live in groups to survive, which Kant admits, but also the result of an innate capacity for reason that commands humans to perfect their faculties, especially the power of rational reflection, which separates them from all other beings on earth and which they can fully develop only within society. Kant maintains, "A human being has a duty to himself to *cultivate* (*cultura*) his natural powers (powers of spirit, mind, and body), as a means to all sorts of possible ends.—He owes it to himself (as a rational being) not to leave idle and, as it were, rusting away the natural predisposition and capacities that his reason can some day use" (Kant 1996a [1797], 565, 6:444, my emphasis). Our innate ability to give ourselves laws, both political and moral laws, can be actualized only in civil society; that is to say, only within a system of political laws in a civil nation-state.

In terms of the moral law, although human beings have a natural capacity for moral discrimination, Kant argues that this primitive power is transformed through a *"pathologically* compelled social union," into "a *moral whole"* through civil society (Kant 2007b [1784], 111, 8:21). Natural, moral, and political right lead to the same results, but through different means; or, more specifically, through different types of laws and different sources of lawmaking (1996c [1795], 334–335, 8:365). Natural laws may compel us to work together to survive, but political laws seal the deal with external laws that we give ourselves and that we give the government the power to enforce. Moral laws have internal coercion that can never become the external coercion of the state. Yet they can compel us to do our political duties as well as our moral ones; indeed, following the laws of the state may be recognized by our practical reason as a moral duty. On Kant's account, one right grows organically out of its predecessor. Ultimately because human beings are unique on earth in that we possess the innate capacity for self-legislation, we are the only *"moral species,"* which becomes the metaphysical ground of both moral and political right.[4] If the seeds of morality are already in nature because of the innate human capacity for reason, we might ask which comes first, ethics or politics? Does the doctrine of right follow from the doctrine of virtue, or vice versa? Kant's answer seems to be that while in terms of chronology, historically, political law necessarily precedes moral law—the doctrine of right precedes the doctrine of virtue—in terms of metaphysics, that is to say logically, however, moral law precedes political law. Kant argues that even devils can be brought into civil society through their own self-interests and the coercion of civil law (Kant 1996c, 335, 8:366); society doesn't need moral laws, just civil laws and the power to enforce them. But, the civil laws, which Kant claims are prefigured in common decency or civility in interpersonal relationships, provide the fertile soil for the cultivation of moral laws.[5] Kant claims, "for it is not the case that a good state constitution is to be expected from inner morality; in the contrary, the good moral education of a people is to be expected from a good state constitution" (1996c, 335–336, 8;366). Historically, then, it is only through a republican constitution that people can develop morally. Still, while civil government is required for us to develop our moral faculties, logically, that moral faculty precedes civil society. For without our moral faculty we would never recognize the authority of political right as anything other than might makes right or the law of the strongest (which in terms of historical or empirical power struggles it is). Kant concludes, "True politics can therefore not take a step without having already paid homage to morals" (1996c, 347, 8:380).

We will return to the question of humans as a moral species; for now the more immediate question is that of the relation of rights to private property, to the possibility of civil law more generally and to our right to possess the earth more specifically. Kant argues that even in the state of nature individuals must have had some primitive sense of right, specifically of property right, or the state could never have been established:

> If no acquisition were cognized as rightful even in a provisional way prior to entering the civil condition, the civil condition itself would be impossible. For in terms of their form, laws concerning what is mine or yours in a state of nature contain the same thing that they prescribe in the civil condition, insofar as the civil condition is thought by pure rational concepts alone. The difference is only that the civil condition provides the conditions under which these laws are out into effect (in keeping with distributive justice). —So if external objects were not even *provisionally* mine or yours in the state of nature, there would also be no duties of right with regard to them and therefore no command to leave the state of nature.
>
> (1996A, 456, 6:312-313)

Whether or not what appears as the circular logic of rights—they are only possible in civil society, but in order for human beings to enter civil society they must already have some notion of rights—as a vicious circle remains to be seen. This "bootstrapping" problem is typical of Kant's thought. And usually he seems to resolve it by separating history from philosophy or physical laws from the laws of metaphysics. Yet, as we will see, when he gets to the question of the right to private property from the provisional rights in nature, the circuit may be especially vicious in that it requires violence, in terms of history, perhaps even the genocide of some humans and some nonhuman animals. Yet, for Kant, it is only through this perpetual war that we come to the possibility of eventual perpetual peace or what he describes as *equilibrium* of political forces or individual state power in an international federation of states (1996c, 325–328, 8:354–357).

Just as individual "savages" in the state of nature are contemptible, "barbarous," "crude," and "brutishly degrading to humanity," and therefore must be forced into civil society (if they cannot be lured in by beautiful women), so too nation-states that do not recognize international law are dangerous and a threat to world peace and must be brought into a united federation of nations even if through war until all human beings on earth are united by one universal law (cf. 1996a 326, 8:354): "In accordance with reason there is only one way that states in relation with one another can leave the lawless condition, which involves nothing but war; it is that, like individual human beings, they give up their savage (lawless) freedom, accommodate themselves to public coercive laws, and so form an (always growing) state of nations (*civitas gentium*) that would finally encompass all the nations of the earth" (1996a, 328, 8:357).

COSMOPOLITANISM TO THE ENDS OF THE EARTH (AND BEYOND)

Before the civil constitution of a state unifies individuals together under one set of laws, their right to property is only provisional. Conclusive property right can only be established by law. Yet, the same applies to nation-states; their property is always only provisional until they subject themselves to international law in a federation of states. Indeed, all private property is provisional until all people and all nations are united under one universal cosmopolitan law that establishes property rights conclusively. Kant argues, "since a state of nature among nations, like a state of nature among individual human beings, is a condition that one ought to leave in order to enter a lawful condition, before this happens any rights of nations, and anything external that is mine or yours which state can acquire or retain by war, are merely provisional. Only in a universal *association of states* (analogous to that by which a people becomes a state) can rights come to hold *conclusively* and a true *condition of peace* come about" (Kant 1996a, 487, 6:350). If all property prior to civil law is merely provisional and, furthermore, if state property is merely provisional in relation to all others, then it follows that property rights are only truly established once cosmopolitan law yields perpetual peace; in turn, it follows that, until then, there are no true or conclusive rights to private property, only provisional rights.[6] So if perpetual peace is impossible, as Kant suggests at the end of *The Metaphysics of Morals*, so too is any true private property. Until there is peace across the globe, there is no legal property. Taking Kant's theory of property to its limit, we see that ultimately there can be no absolute principled justification for private property, whether it belongs to the individual or to the state.

Furthermore, Kant seems to require that the universal cosmopolitan law that unites all citizens of the earth (*Weltbürgerrecht*) extend to the entire human race: "there must be some original acquisition or other of what is external, since not all acquisition can be derived. So this problem cannot be abandoned as insoluble and intrinsically impossible. But even if it is solved through the original contract, such acquisition will always remain only provisional unless this contract extends to *the entire human race (Menschengattung)*" (1996a, 418, 6:267; my emphasis). This claim implies that until the entire human race is united under one civil law there is no true right to private property. In other words, until the entire human race is bound by the same law, all property is provisional. Again, at its limit, Kant's theory of property collapses under the weight of the demand for universal principled justification. On the one hand, private property is justified by the original common possession of the surface of the earth by all human beings. And, on the other hand, true private property can be established only once all human beings submit to a common law. Public right, then, necessitates a global thinking that at once underpins Kant's theory of property while at the same time undermining it.

At this point, we might wonder what Kant means by "the entire human race." *Menschengattung*, which is variously translated as "human race," "human species," "human beings," and "people" connotes the entire species of human beings. In historical terms, the entire human species includes past, present, and even future human individuals. If Kant is appealing to the human species in these terms, then private property is absolutely impossible. Once again, he must be appealing to a principle or law of reason that unites all human beings across time, a principle perhaps that constitutes human beings as a species. And, as we have seen, the notion of human beings as a species is central to both Kant's political and moral philosophy.

In addition, if we take Kant's conjectures on history seriously, then we know that the human faculty of reason, although innate, develops through experience and reaches its perfection not through the experience of individuals but through the cumulative experience of the species. Kant makes it clear that man's natural capacities of reason are such that they can only be developed in the species and not in the individual (Kant 2007b [1784],

109, 8:18). He claims that it may take an ""immense series of generations."" for our species to develop into the rational one that nature intended (2007b [1784], 110, 8:19); my emphasis). He goes so far as to say that individuals are not only unconscious of this development but also uninterested in it (2007b [1784], 108, 8:17). If, as Kant maintains, human beings are the one animal species intended to have reason and develop it through the evolution of their species because as a class they are, as he says, "mortal as individuals but immortal as a species,"[7] we must wonder whether the a priori rational law that unites all human beings will develop through the evolution of the species so that only at the end of such time as it is perfected will we recognize, and choose to live by, it and only then will private property be legitimately justified. In other words, the concept of private property, insofar as it requires the perfection of reason in the human species, is possible only after the end of time.

Perhaps this is why Kant turns away from the temporal arguments over original acquisition based on first possession and toward the spatial argument based on common possession of the earth's surface. Kant is troubled by, yet attempts to justify, the retroactive temporality of original acquisition through which the provisional right to property in the state of nature is eventually "favored" (*Gunst*) by civil law within the nation-state. By turning to the spatial metaphor of the limited surface of the earth, the justification for private property becomes a problem of distribution rather than original access. In other words, the limited surface of the earth is divisible because it is finite. And because it is divisible, private property is possible. Since Kant is committed to fundamental principles of freedom and equality, the distribution of property must be based on these principles.[8] This is also why Kant imagines the possession of the earth's surface as "collectively universal, i.e., as the common possession of the human species to which corresponds an objectively united will or will that is to be united," which is what determines the possibility of equal distribution. In other words, the equal distribution of private property is based on the united will of all mankind.[9] From this, Jeffrey Edwards deduces what he calls a principle of "material equality," which would limit the amount of the earth's surface that any individual could rightfully own. He concludes that Kant's theory of original acquisition must have a formal principle of material equality in order to provide a normative foundation for property law. If this is so, then Kant's theory of private property undermines capitalism as the right to accumulate property.[10] Or, as Edwards puts it: "Kant's theory of original acquisition must end up destroying the normative basis of capitalist property relations if it is to be consistent with his conception of laws of right as laws of external freedom."[11]

The fact that Kant bases both provisional and definitive property rights on all human beings' equal possession of the finite surface of the earth means that, ultimately, the so-called original contract is grounded on the fact that we all have to live on the same planet.[12] The original contract, then, is already in effect in the state of nature insofar as we tacitly agree to live together on the planet. But it becomes a rational choice only within civil society. So the original contract is based on our tacit agreement to inhabit this globe together insofar as (as of yet) we have no other choice. Yet it is precisely our inability to choose what planet we inhabit that gives rise to our choice, and the significance of that choice, when it comes to private property.

Kant describes it thus:

> All human beings are originally (i.e., prior to any act of choice that establishes a right) in a possession of land that is in conformity with right, that is, they have a right to be wherever nature or chance (apart from their will) has placed them. This kind of possession—which is to be distinguished from residence, a chosen and therefore an acquired lasting possession—is a possession in common because the spherical surface of the earth unites all the places on its surface; for if its surface were an unbounded plane, people could be so dispersed on it that they would not come into any community with one another, and community would not then be a necessary result of their existence on earth.—The possession by all human beings on the earth which precedes any acts of theirs that would establish rights (as constituted by nature itself) is an original possession in common, the concept of which is not empirical and dependent upon temporal conditions, like that of a supposed primitive possession in common (communion *primaeva*), which can never be proved. Original possession in common is, rather, a practical rational concept which contains a priori the principle in accordance with which alone people can use a place on the earth in accordance with principles of right.
>
> (KANT 1996A, 414–15; 6:262)

Kant distinguishes original possession, or living on the land (what he calls "holding" it), from lasting possession, or willfully choosing a residence (what he calls "having" it) (1996a, 414; 6:262). It seems that all human beings have an equal right to own land because they will it. But then either Kant is back to the temporal problem of original acquisition or the spatial distribution of that land requires material equality—or at least equal opportunity to acquire it.[13]

DECLARING "THIS IS MINE!"

One of the reasons Kant argues that originally all humans possessed the surface of the earth is because private property only makes sense if it was acquired and prior to its acquisition by a particular person did not belong to that person but to another. Yet if it belonged to another particular person, then the person acquiring it would be infringing on that person's freedom. So it must have belonged to everyone in common, although not in the sense of possession as ownership or having, but only in the sense of possession as occupying or holding. Only if it belonged to everyone in common can parts of it be separated off for private use—although sometimes Kant says that only if it belongs to no one can someone claim it as their own (cf. Kant 1996a, 411, 6:258). How this happens, of course, takes us back to the problem of original acquisition. At this point, rather than return to the bootstrapping problem of retroactive temporality, we turn our focus to another crucial aspect of original acquisition as Kant describes it, namely the role of signs or the declaration "this is mine." In his discussion of how to acquire something "external," Kant identifies three stages of acquisition: 1. *Apprehension*, that is to say, physically taking possession of it; 2. *Giving a sign (declaratio)* of my possession of this object and my act of choice to exclude everyone else from it. 3. *Appropriation*, as the act of a general will that binds everyone else to agree with my choice (1996a, 411, 6:259). The transition from mere physical possession in the state of nature to true or rightful possession circumscribed by civil law necessitates a sign or declaration that "this is mine." But, given that living on the land is not enough to establish acquisition, what counts as a sign that I want this property for my own use at the exclusion of others? It is clear that what needs to be added to provisional property in the state of nature is civil law based on the act of a general will that binds everyone universally. But it is not so clear what constitutes the sign or declaration of a particular will to possess a particular piece of land. Certainly, the ability to protect the land with weapons is a powerful sign of the intent to possess it. Indeed, Kant suggests not only that possession requires the ability to "protect" and defend the land but also that the extent of possession is governed, if not determined, by it (Kant 1996a, 416, 6:265).[14] For example, a country owns the seas to the extent that it can defend the shore (1996a, 420, 6:269).[15] Of course, Kant was writing before the advent of nuclear weapons and rocket launchers whereby one country might destroy any other.

Kant also argues that developing the land, while not proof of ownership, does count as a sign (1996a, 417, 6:265).[16] In answer to the question of whether developing the land is necessary for acquisition, Kant answers: "No . . . developing land is nothing more than an external sign of taking possession, for which many other signs that cost less effort can be substituted" (1996a, 417, 6:265). My concern is with these "many other signs" and how are they interpreted. For it seems that in order for any sign or declaration of possession to be effective, it must be recognized as such by others. In other words, it doesn't work to merely say to oneself *this is mine;* someone else must also hear the declaration and understand it or "read" and interpret the sign. While cannon fire is difficult to misinterpret, it is not the only sign of possession. We know from history that the Europeans who "discovered" so-called uninhabited lands too often massacred the natives who lived there as a sign of their will to possess them; indeed, sometimes they declared ownership not only of the land but also of the natives. Kant says that we must repudiate this means of acquiring land, even if the ends—bringing the savage natives into civilization—is a good one (1996a, 418, 6:266). The fact that some people interpret land as uninhabited even though there are native inhabitants raises the question of what counts as habitation. Even before we get to the question of signs of a particular will, habitation itself is open to interpretation.

While Kant does not condone forcibly taking land away from natives, even if he can see how it might be for the betterment of humankind, he is clear that animals have no claims whatsoever to possession of the earth. Indeed, animals are things like land that can be possessed by human beings.[17] Like any other property, we can lay claim to animals by apprehending them, declaring them our own, and then appropriating them via the general will of possession, which, as we have seen, is ultimately founded on the finite surface of the earth. Just as the earth is ours, so too are its plants and animals.

Plants and animals are the accidents of the substance that is the land (more on this distinction later). For Kant, then, only humans inhabit the land. Animals do not count as inhabitants, at least not for the sake of determining property rights. Whatever we think of Kant's proclamations about animals being things or property for our use, the distinction between humans and nonhuman animals comes to bear on questions of habitation in the cases just mentioned where European explorers claimed to discover new, as of yet uninhabited, lands. For, the only way in which they could have considered these lands undiscovered and uninhabited is if they considered the natives to be animals rather than human beings. In other words, the question of what is a human being—or, more to the point, what is recognized as human—has everything to do with whether or not land is considered inhabited. The habitation of native peoples and their various tools and strategies for defending "their land" were not recognized as signs or declarations of first possession by European imperialists. If we cannot be sure that we can read a sign so forcefully presented by another of our species, how can we know whether or not other species are capable of giving signs or expressing their will to possess land? And if they do, are they not also inhabitants?

For example, what are we to make of the fact that young male African elephants whose parents have been killed by humans are beginning to attack human villages for the first time (see Bradshaw 2004 and Siebert 2006)? Can we interpret the violent acts of these gangs of traumatized adolescent males not only as revenge but also as signs of their willingness to possess their territory at the exclusion of others, particularly human others? And what about the two young male lions who halted the construction of a British railroad bridge over the Tsavo River in East Africa for almost a year in 1898 when they killed railroad workers nightly at different camps along the river? They killed at least 35 workers and perhaps as many as 135. Reportedly, the nightly disappearance of workers was a mystery to some, especially the British officer in charge. The lions were able to move through the thorn fences constructed to keep them out, and some said that the lions were able to predict the movements of Colonel Patterson who was overseeing the railroad construction (Patterson 2004, Patterson 1907).[18] Eventually, Patterson tracked and killed the lions. Were there lions asserting their rights to this land? Were they defending it against invaders and claiming it as their own? And can we be so sure that among lion-kind they had not formed a united general will that they, the kings of the jungle, if not members of the kingdom of ends, have property rights?

The question is not only what constitutes a sign or a declaration but also who is authorized to interpret those signs? Who is the judge that arbitrates what counts as a declaration, especially within the state of nature before the existence of civil judges or civilization? And, from the standpoint of civilization, anything left in the wild is incapable of signs, incapable of inhabiting, and incapable of possession. The transition from the state of nature to civilized society through signs and declarations once again already presupposes a backdated judgment of an authorized, civilized judge who arbitrates rights retroactively even before the existence of (civil) rights, properly speaking. As the aforementioned cases of conquering uninhabited lands shows, the question of who or what counts as human is not just an empirical question but also, moreover, a metaphysical one. And the metaphysical divide between the human species and other species on earth is what is used to justify our possession of the surface of this planet upon which we all live and depend.

THE POROUS SURFACE OF THE EARTH: SUBSTANCE VERSUS ACCIDENTS

Our dependence upon the earth and the plants and animals with which we share the planet raises another significant question for Kant's theory of property. Crucial to Kant's theory is the ability to separate external possessions from our own bodies, to separate our bodies from other bodies. Furthermore, given that so much of Kant's doctrine of right is about commerce, he takes pains to separate the products of the land from the land itself such that one can sell the products while still owning the land. While it seems obvious that we are separated from the plants and animals and other stuff on the surface of the earth, in an important sense we are also all inherently interdependent. Of course Kant did not have the benefit of the science of ecology or theories of environmental interdependence and ecosystems. But, his cosmopolitanism is based in part on the fact that what happens in one part of the globe affects what happens in another. He was keenly aware of our global interconnectedness and therefore the need for cosmopolitan rights. Political right requires us—or more accurately, *should* require us—not to wrong our neighbors, particularly in terms of their property. Contemporary concerns with air and water pollution, among others, make it more difficult to contain the harms done to another's property and separate that property from one's own. While Kant defines land as inhabitable ground, he neglects the other elements that make the earth habitable such as air and water (Kant 1996a, 414, 6:261). These gaseous and fluid elements are always in excess of property rights founded on the solid ground of the earth's surface.

Yet is the earth's surface so solid and unchanging that it can found universal property law? Certainly, in terms of the geometry of a sphere, the characteristics upon which Kant bases his claims are unchanging: namely that a sphere has limited surface and that all places on its surface are connected or united by its shape. The identification between the earth's surface and "land," however, allows for more slippage than Kant's theory can admit. Just as Kant reduces the earth to geometry, so too land can only be an abstract concept. Of course the realm of the abstract and a priori is the realm of philosophy. Yet, given that Kant's philosophy also relies on relationships between people and the exchange of the earth's resources, the earth's fluid ecosystem seeps in. Indeed, he defines property in terms of relations between people rather than relations between people and things or land; after all, property relations are agreements between people and not agreements between people and the land itself.[19] For Kant says that if it were, the land would refuse "possession of itself to anyone" (1996a, 405, 6:250n). At this point we might ask what signs or declarations might the land or earth itself make in its own defense against possession by others? Moreover, the fact that Kant's theory depends so heavily upon the *earth's surface* might make us wonder about the possessive relationship suggested between the earth and its *own surface*—the surface belonging to the earth. We might ask, how is the earth's possession of its own surface different from our original common possession of it? What does possession mean in each of these cases? And, furthermore, does the earth possess its surface as an accident or as substance?

Kant did not ground property rights merely on the fact that the earth is a globe—that is, on its geometric shape—but also on the fact that it has limited surface that can be distributed among human beings. He is clear to separate the earth from its surface, especially once we consider writings in which Kant discusses the molten core of the earth and the raging heat underneath its crust that can burst forth and cause earthquakes and tidal waves. The earth's surface is porous and subject to outbursts from the earth's molten core. Its surface, as Kant notes, is necessarily irregular (that is to say not flat but mountainous) and changing. Moreover, when push comes to shove—or should we say shovel?—it is difficult to determine where the earth's surface stops and the earth itself starts. Even Kant's legalistic discussion of the distinction between the products of the land, what he calls accidents, from the land itself, or what he calls substance, is problematic in this regard (Kant 1996a, 414, 6:262). Given the place of original acquisition in his theory of property, the distinction between the land itself and its products is essential. For, as he maintains, the use or development of products of the land does not alone constitute possession. And, in order for commercial exchange between people or nations, products must be separable from the land upon which they grow or lay. Yet, again, we might ask where the accidents stop and the substance starts. At what point do dirt and rocks stop being accidents that we can sell and start being part of the substance of the land? And what about the natural gases and fossil fuels beneath the earth's surface? Are they accidents or substance? Again, Kant's answer would involve the concepts rather than the material (since material is accidental to philosophy, while concepts are its true substance). Yet, given changing conceptions of land, products, ownership, and relations between them, as Kant is well aware, property disputes are practical affairs in which, more often than not, might makes right. And while Kant accepts that in the beginning it was necessary provisionally for might to make right, now, and in the future, only law makes right. Nevertheless, the problem of how to separate the soil, rocks, water, animals, plants, trees, and other accidents from the substance of either land or the earth is both a material and a conceptual one that Kant's theory does not solve. Indeed, Kant's theory cannot account for the fact that the surface of the earth is constituted through earth's biosphere or life. In the words of the Russian scientist Vladimir Vernadsky, one of the first to use the term *biosphere*, "Without life, the crustal mechanism of the Earth would not exist" (1998, 58).[20] The earth's surface is a dynamic living force, a network of relationships, which is lost when it is reduced to mere geometry or mapmaking.

THE EARTH'S INHOSPITABLE HOSPITALITY

The earth is hospitable to human beings because it provides us with these accidents, which are none other than its resources. In "Perpetual Peace,"

Kant marvels at the care that nature takes so that "people should be able to live in all regions of the earth," even the most "inhospitable" (*Unwirtbarste*) (Kant 1996c [1795], 332, 8:363). Kant lists some of these wonders:

> That moss grows even in the cold wastes around the Artic Ocean, which the reindeer can scrape from under the snow in order to be the nourishment, or also the draft animal, for the Ostiaks or Samoyeds; or that the sandy wastes contain salt for the camel, which seems as if created for traveling in them, so not to leave them unused, is already wonderful. But the end shines forth more clearly when we see that on the shore of the Artic Ocean, there are, besides fur-bearing animals, also seals, walruses, and whales, whose flesh gives the inhabitants food and whose blubber gives them warmth. But nature's foresight arouses the most wonder by the driftwood it brings to these barren regions (without anyone knowing exactly where it comes from), without which material they could make neither their boats and weapons nor their huts to live in; there they have enough to do warring against animals, so that they live peaceably among themselves.
>
> [1996C [1795], 333, 8:363]

Even the most inhospitable places on earth can be inhabited thanks to the wonders of nature, which provides for our needs in the most unlikely ways. Eventually, we come to provide for each other through trade agreements that bring products from one part of the earth to another. These agreements are bought at the cost of wars that result from fighting over the earth's resources. Yet the earth's inhospitality seems to force subsequent peace since we need each other's help to survive such harsh conditions. Without commerce, or what Kant calls the dynamic community, we could not—or at least would not—live everywhere on earth. The earth's inhospitality necessitates the right to hospitality, which makes cosmopolitan right possible. Human beings must be hospitable to each other when the earth is not hospitable to them. Thus nature provides for us through the earth's resources in the most wonderful ways, which are evidence of a certain natural hospitality, yet harsh climates and natural disasters such as earthquakes and tidal waves are evidence of nature's inhospitality.

In the face of the inhospitality of nature, we must come together as a cosmopolitan community. Natural disasters, in particular, can bring people together. For example, at the end of his second essay on earthquakes, Kant suggests a noble prince would stop warring on a country that was the victim of a natural disaster such as an earthquake and that such a prince would be a "gift to the peoples of the earth" (Kant 2012b [1754], 364, 1:461). Furthermore, he argues not only that there are benefits to earthquakes but also that earthquakes are the flip side, so to speak, of what is most advantageous about our life on earth, including healing warm baths (2012b [1754], 360, 1:456).[21] Kant concludes that there are so many advantages to the heat beneath the earth's surface that we should approach even earthquakes with gratitude for what the earth provides.

THE BODY'S INHOSPITABLE HOSPITALITY

What we might call the earth's *inhospitable hospitality*, then, gives rise to what Kant calls man's *asocial sociability*. We need to live among other people; but we are constantly fighting over the earth's resources laying claim to them as our own private property, which can be guaranteed only in times of peace by universally recognized laws. In *Anthropology from a Pragmatic Point of View* Kant states, "the characteristic of the human species is this: that nature has planted in it the *seed* of discord, and has willed that its own reason bring concord out of this, or at least the constant approximation to it" (Kant 2006, 226, 7:322; my emphasis).[22] Just as we should be grateful for the earth's inhospitable hospitality, which eventually leads to political union, Kant claims that we should be grateful for our own asocial sociability since it leads to the development of our faculty of reason and the progress of mankind: "Thanks be to nature, therefore, for the incompatibility, for the spiteful competitive vanity, for the insatiable desire to possess or even to dominate! For without them all the excellent natural predispositions in humanity would eternally slumber undeveloped" (Kant 2007b, 112, 8:21). Without asocial sociability, human talents "remain hidden in their germs" and men would be "as good-natured as the sheep" they tend (2007b, 112, 8:21). In this case Kant insists that Nature knows best and has endowed us with asociality so that we can reach our higher purpose, beyond what is possible for docile sheep.

So, too, the inhospitable hospitality of the earth is mirrored by the inhospitable hospitality of our bodies to reason, which makes us human and distinguishes us from all other species. Just as the earth is hospitable to human life, our bodies are hospitable to reason and other faculties. Just as plants, animals and humans grow and mature on the surface of the earth, so too reason grows and matures in the human. In his earliest writings, before figuring human beings as endowed with reason or humanity, Kant talks of the invisible spirit *inhabiting* the body (Kant 2012c, 298, 1:356). The human species is distinguished from all others in that our bodies are inhabited by reason. Throughout his writings, it is clear to Kant that, among earthlings, only the human body is hospitable to reason and to humanity. Other animal bodies are presumably uninhabitable and inhospitable to reason. Among our fellow creatures on earth, only we possess reason as a property or faculty of our minds and only we possess humanity as a result of our rationality. Yet, if we take seriously Kant's remarks on natural science and anthropology, our capacity for reason is governed, if not determined, by our bodies and by the bodies that are hospitable or inhospitable to our own, such as the planet Earth, on the one hand, and bodies that we ingest and that are hospitable, so to speak, to our substance, on the other. In "Universal History," Kant says that the mind and reason are dependent upon the physical matter of the body; and all of our concepts and ideas are not only originally stimulated through the body but also the constitution of our matter determines our capacity to think (2012c, 298, 1:355–356).[23]

Indeed, in his conjectural history Kant imagines that reason first developed in relation to food and man's curiosity about what may or may not be edible. There he describes the development of reason from instinct and inclination, specifically in terms of human experimentation with taste and smell. He suggests that humans develop both their imaginative capacities and rational capacities in relation to these early experiments with food. Furthermore, reason appears on the scene to interrupt man's blissful eating: "So long as inexperienced man obeyed this call of nature, his lot was a happy one. But reason soon made its presence felt and sought to extend his knowledge of foodstuffs beyond the bounds of instinct" (2003 [1786], 223). The capacity of reason not only develops in relation to eating but also transforms man's natural food intake into knowledge of cause and effect, what foods produce what consequences. For Kant, this is the recipe for human development from an animal state to a human one.

Still, Kant is at pains to reconcile our animal nature and our rational one, we could say that the human body exhibits an inhospitable hospitality to reason. Just as human beings are warring on the planetary body Earth, so too reason is warring within the human body. The doctrine of right is a story of external war that eventually leads to peace, while the doctrine of virtue is about internal war that eventually leads to peace. And both reach peace through the evolution of the human species on earth; outer peace makes possible inner peace. And yet reason in both right and virtue are defined by war, war between nation-states or peoples and war between reason and inclination or mind and body. Moreover, the transition from animal to human, one that, it seems, is still underway, is filled with anxiety that is definitive of human existence. Again speaking of the experimentation with foodstuffs that leads to rational choice, Kant claims that humans, unlike other animals, discovered the ability to choose what to eat from a great variety, but that this realization, which gave them superiority over other animals, also brought with it anxiety and fear over such decisions.[24] He could no longer rely on his instincts and in itself his freedom gave him no guidance as to what he should choose. Kant says, "He stood, as it were, on the *edge of an abyss*. For whereas instinct had hitherto directed him towards individual objects of his desire, an *infinite range of objects* now opened up, and he did not yet know how to choose between them. Yet now that he had tasted this state of freedom, it was impossible for him to return to a state of servitude under the rule of instinct" (Kant 2003, 224). We are a hybrid species, reason dependent upon an animal body. Or, as Kant puts it in his early scientific essay on the place of earth in the solar system, "intellectual abilities have a necessary dependence on the material of the machine they inhabit" (2012c, 301, 1:359). In this same essay, "Universal Natural History and Theory of the Heavens," he compares human beings to plants who "absorb sap," grow, reproduce, become old and die" (298, 1:356). As Elaine Miller deftly demonstrates in *The Vegetative Soul*, Kant's writings are full of plant metaphors through which he explains the way in which our faculty of reason operates as part of, yet separated from, nature. For example, in "Idea for a Universal History" he likens enlightenment to the implanting of germs: "then nature perhaps needs an immense series of generations, each of which transmits its enlightenment to the next, in order to finally propel its germs in our species to that stage of development which is completely suited to its aim" (2007b, 110, 8:19). There, he also compares man to a tree when he says, "Nothing straight can be constructed from such warped wood as that which man is made of" (2007b [1784], 46). These are just a couple of examples of numerous plant metaphors throughout Kant's writings.

Thus, even while explaining how human reason transcends nature, Kant relies on metaphors from nature. In addition, central concepts of political philosophy are also contained withinhis philosophy of nature. Combined with remarks from Kant's *Anthropology*, his early essays in *Natural Science* suggest that the physical characteristics of the earth govern, if not determine, both the political and moral characteristics of its peoples. Specifically, the notions of *causality, equilibrium,* and *maturity* play important roles in both Kant's philosophy of natural science and in his political philosophy. In *The Metaphysics of Morals* and in his conjectural histories, Kant attempts to identify the causes of political alliance, private property, and public right. He does so by appealing to physical proximity and natural forces that eventually become or are transformed into rational laws. More specifically, when examining how causality works in some of his writings on natural science, we see that he identifies a logic of good and evil or life and death that inheres in one and the same cause. In other words, the same event or characteristic causes both good and bad effects. For example earthquakes and volcanoes cause death and destruction, but they also release beneficial heat from beneath the surface of the earth, for which we should be grateful. So, too, the very causes that bring a body to maturity, whether it is a human body or the earth itself, also cause it to decay and perish (Kant 2012b [1754], 170).

This logic whereby opposing effects are the result of the same causes is also manifest in Kant's discussion of politics and morals. For example, he maintains that war and peace have the same cause, the fact that we share the limited surface of the earth and its resources. War, he suggests, is necessary for both political alliance and for eventual peace.[25] In addition, he proposes that the very causes of our rationality and thus our morality are also the causes of our irrationality and our capacity for evil. The good and the bad are two sides of the same coin, so to speak. The lesson he draws from this dual-effect causality is that we should be grateful for both the good and the bad. Ultimately, Kant suggests that natural disasters and other evils are pedagogical tools from which we can learn. And one of the most important lessons is that the "husbandry of earth is entrusted to man" (2012a, 340, 1:431).

The best that we can hope for is equilibrium between forces, both in terms of political struggles or war and natural pressures or natural disasters. The Earth, claims Kant, emerged from chaos and eventually attained equilibrium, and the ensuing order again displays the fortunes of Providence for which we should feel grateful. Kant suggests that the fruits of the earth are the result of this equilibrium and order: "After nature attained this state of order and established itself in this condition, all the elements on the surface of the Earth were in a condition of equilibrium. Fertility spread its riches in every direction, and the Earth was fresh in the flower of its strength, or if, I may so put it, it was in its age of manhood" (2012b [1754], 171, 1:200).

Kant also describes political alliance born out of the chaos of war in terms of equilibrium. He says that wars compel our species to discover a law of *equilibrium* to regulate the essentially healthy hostility which prevails among states and is produced by their freedom" (2007b [1754], 115, 8:26). In both *The Metaphysics of Morals* and "Perpetual Peace," Kant suggests that peace is the result of equilibrium of opposing forces (1996a, 488, 6:350, 1996c, 336, 8:367). For Kant, peace between nations or peoples and peace between reason and matter or bodies is the result of this equilibrium, which is the product of what he calls *maturity*. Just as some areas of the earth are "immature," so some peoples of the earth are "immature" (cf. Kant 2012b [1754], 171, 1:200). Reason, like the earth and races, must age and mature in order to come into its own perfection. As Derrida points out, however, Kant isn't always consistent; Kant claims that human reason is perfectible through maturation and yet he speculates that if human beings lived to be eight hundred years old they would turn to vice out of boredom (Derrida 2011, 274). Of course, when Kant discusses human perfectibility and maturity, he is speaking of the species and not individuals.

Kant is also inconsistent when it comes to whether or not soil and landscape govern or determine the political and moral character of peoples. Although in places he clearly insists that they do not, he also proposes that they do. For example, in spite of his claim in his *Anthropology* that "climate and soil" cannot "furnish the key" to the character of a people, he is also famous for remarks such as the following: on the difference between the French and the English, he says, "[that] English [is] the most widely used language of commerce, especially among business people, probably lies in the difference in their continental and insular situation" (2006, 214, 7:312); while "the Italian unites French vivacity (gaiety) with Spanish seriousness (tenacity), and his aesthetic character is a taste that is linked with affect; just as the view from his Alps down into the charming valleys presents matter for courage on the one hand and quiet enjoyment on the other" (2006, 218, 7:316). He claims that the German "is too cosmopolitan to be deeply attached to his homeland. However, in his own country he is more hospitable to foreigners than any other nation" (2006, 221, 7:318).

In his *Physical Geography* Kant makes the link between landscape or soil and political economy clear when in the announcement of the program for his lectures he states, "The condition of states will rather be considered in relation to what is more constant and which contains the more remote ground of those accidental causes, namely, the situation of their countries, the nature of their products, customs, industry, trade and population" (quoted in Elden 2012, 7). There, he argues that moral geography or the moral customs of peoples, political geography or the governments of states, mercantile geography or commerce and economies, and theological geography or religions are all grounded in physical geography. For example, he claims that political systems are founded on the characteristics of the land and that political geography is "founded on physical geography" (2012a, 452, 9:164).[26] He also maintains that economies are based on the characteristics of physical geography insofar as trade is based on the natural resources of countries that exchange what they have for what they need (2012a, 452, 9:165).[27] Even religion is the result of the soil: "Since theological principles frequently undergo fundamental changes according to difference of soil, essential information will be provided concerning this as well" (2012a 452, 9:165). In all areas of our social existence, moral and political, economic and religious, we are governed if not determined by the characteristics of the surface of the earth. In the deepest sense we are peoples of the earth.

We are so intimately and essentially connected to the earth that Kant claims, "to know the human being according to his species as an earthly being endowed with reason especially deserves to be called *knowledge of the world*, even though he constitutes only one part of the creatures on earth" (231, 7:119). And yet we can only know ourselves by comparison with those other creatures on earth. At least in terms of the *Anthropology*, human beings are defined in terms of, and against, other inhabitants of earth.[28] In order to get the proper perspective on human beings as inhabitants of earth, however, Kant also compares us to extraterrestrial aliens from outer space. As he does so often, Kant describes our place in the cosmos relative to aliens from other planets (e.g., Kant 2007b [1784], 47*n*).[29]

Kant's is a terrestrial perspective—we are the ends of nature on earth, but not in the universe. Indeed, while Kant concludes that human beings are the most perfect species on earth because of our superior intellectual (and therefore moral) powers, ultimately our place among rational beings is uncertain because we do not know our place in the universe: "The highest species concept may be that of a terrestrial rational being. However we will not be able to name its character because we have no knowledge of non-terrestrial rational beings that would enable us to indicate their characteristic property and so to characterize this terrestrial being among rational beings in general. It seems, therefore, that the problem of indicating the character of the human species is absolutely insoluble, because the solution would have to be made through experience by means of the comparison of two species of rational being, but experience does not offer us this" (2006, 225, 7:321). Given that we have experience of only one rational species, namely ourselves, we do not truly know what it means to be a rational species. Only if we could compare the rationality or reason of another species could we begin to understand the capacity for reason. Yet Kant imagines that one day we may travel to other planets, perhaps even reside there, and that "the satellites orbiting around Jupiter" may "light our way in the future" (2012c, 307, 1:367). And perhaps in the future we will encounter other rational beings through whom we may learn about ourselves. Thus only through an alien perspective can we truly see ourselves.

My earlier comparison of first contact from Star Trek may not be far off the mark. For Kant speculates that all planets are inhabited now or will be in the future. And within the Kantian universe we can imagine that all rational beings, terrestrial and extraterrestrial, earthlings and the inhabitants of other planets, must form a federation of planets to ensure not just a *cosmopolitan* peace but also a *cosmological* peace that will spread across the entire galaxy, which does not belong to us. While human beings may originally enjoy common possession of the earth's surface, the same cannot be said of other planets. And it is here, when Kant imagines the earth and earthlings from the vantage point of the universe, that we see ourselves humbled (cf. Kant 2012c, 307, 1:368). Even as he imagines human beings leaving their terrestrial home, he admits that we cannot survive on other planets any more than their inhabitants could survive on earth. These other planets are not habitable; they are not hospitable to us. As the 1997 *Pathfinder* Mars time experiment suggests, even the slightest shift in earth's rotation on its axis or alternation of our sense of night and day is intolerable (in order to track the movements of the *Pathfinder* on Mars, scientists lived on Martian time, but within a month the mere thirty-nine minutes and thirty-five seconds longer per day became unbearable).

In sum, Kant's cosmological view both grounds and is in excess of his theory of property. Only from the vantage point of extraterrestrials do we see ourselves as earthlings necessarily inhabiting the same planet with its limited surface. On Kant's theory the truth of property can only be established once the entire human species comes under universal laws of public right. And while the human species qua species is distinguished from other terrestrial species by its capacity for reason, extraterrestrial beings may share this capacity. Insofar as that is possible, then the truth of property, and our right to the earth as our own, is established on when universal laws of right founded on principles of reason extend to the entire universe of rational beings. Not only cosmopolitan peace but also cosmological peace is necessary to establish private property rights with any certainty. Like peace, then, private property itself may be a regulative ideal that can never be attained. Private property is always in excess of our common possession of the earth's surface upon which it is grounded. And the earth and our relationship to it, and dependence upon it, are in excess of Kant's theory of property.

We are earthbound creatures formed by our relationship to this planet and all of its inhabitants. We belong to the earth. And, in a sense, it belongs to us—perhaps not in the Kantian sense of possession, but rather in the more archaic sense of accompanying.[30] The earth belongs to human beings and we belong to the earth, not as possessions but as companions. We belong to each other, but perhaps only temporarily, particularly if we do not learn to share the resources of the earth with its other inhabitants, human and otherwise. The earth's inhabitants do not belong to us except insofar as we all belong to the earth. And earth is not our possession, even if it is our home.

3

PLURALITY AS THE LAW OF THE EARTH

Arendt

> Human beings in the true sense of the term can exist only where there is a world, and there can be a world in the true sense of the term only where the plurality of the human race is more than a simple multiplication of a single species.
> —HANNAH ARENDT, *THE PROMISE OF POLITICS*

Like Kant's doctrine of right, which is circumscribed by the spherical shape of the earth, Arendt's notion of politics is similarly based on "the limited space for the movement of men" determined by the surface of the earth (Arendt 1958a, 52). As we have seen in our analysis of Kant, however, the earth remains in excess of Kant's doctrine of right insofar as it is based on private property. The earth cannot be contained within his theory of property. Indeed, the earth with all of its inhabitants exceeds his notion of the right to hospitality and his limited cosmopolitanism as guarantees for peace, or, more accurately, we might say Kant's extraterrestrial relation to the earth makes his cosmopolitanism itself exceed the boundaries that he tries to set for it insofar as ultimately it requires that all rational beings, be they human or not, earthlings or not, throughout all of time and all of space, be united in a general will.

Whereas Kant imagines a distant future in which the human species will have evolved into a peaceful race of citizens of the world, Arendt, respects the need for international law, even while she abhors cosmopolitanism, which she imagines as a nightmarish tyranny of one government over the "whole earth" (Arendt 1968a, 81). Like Kant, Arendt is more comfortable with a federation of nation-states than a world government. Furthermore, she warns against thinking of human beings as a *species* when talking about politics. Although, unlike Kant, Arendt makes the earth a central theme in her writings, the figure of the earth, and the work that it does in her philosophy, exceeds and outstrips the limitations of her political theory. Taking up the related themes of earth and world, home and statelessness, and witnessing and imagination helps to develop an ethics of earth conservation through analyzing, extending, and sometimes deconstructing those pairs in Arendt's political philosophy. Analyzing Arendt's comments about war and world, we can extend Arendt's insistence on cohabitation and sharing the earth as the ground of politics into the sphere of ethics by putting pressure on her distinction between *just* war and *total* war. Finally, expanding Arendt's notion of *amor mundi*, ultimately politics must be grounded not only on love of world, but also on love of earth, through which Arendt's care for the world is extended to the Earth and all of its inhabitants. For, even if we do not share a world with others radically different from ourselves, we do share the earth.

Arendt's notion of plurality comes up against its limit when we consider the "right" to cohabit the earth. Our coexistence on earth is prior to Arendt's "right to have rights." Whatever the problems with Arendt's theory of rights, taken to its limits, her insistence on plurality as "the law of the earth" entails embracing not only individual and cultural differences but also species difference and biodiversity. Moreover, the question of where to draw the line in terms of what constitutes plurality or a world becomes the sore spot in Arendt's endorsement of just war. Engaging with Arendt's arguments against what she calls "total war," or genocide, as we did with Kant's theory of property, pushes her theory to its limits by questioning what counts as plurality or "a people." Whereas Kant's justification for private property ultimately necessitates implicit acceptance by a general will that unites all rational beings not only on earth but also across the universe and time and space, which taken to its limit makes any justification for private property untenable, Arendt's justification for war between nation-states ultimately rests on the destruction of any human being as the annihilation of a world, and perhaps even of "a people," such that any justification for war becomes untenable. In other words, just as, once we follow Kant's expanding universe, any defense of private property becomes questionable, once we follow Arendt's shrinking world from "a people," to a nation-state, to a family or perhaps a couple, to that singular newborn conjured by natality, any justification for killing even one human being becomes questionable. And, if we extend her notion of cohabitation and plurality to other species, then killing even one living being destroys not only a world but also the possibility of the world. For, "total war" is what Arendt calls a "mortal sin" precisely because it not only destroys a world but also threatens to destroy the very structure of world.

Several scholars have used Arendt's theories, particularly those of earth and world, along with their alienation, to discuss environmental ethics.[1] This chapter takes another turn toward the role of Earth as both a limit concept and a concept of limit that has the potential not only to transform our thinking about political responsibility but also to ground politics in an ethical obligation conceptually, if not chronologically, prior to the possibility of politics. Arendt's philosophy is essentially about limits and borders.[2] It is about the borders between concepts, and it is about the limitations—even dangers—of importing a concept from one realm of life into another realm. It is about maintaining borders between nations and the dangers of abolishing nation-states, while, at the same time, it is about limiting state sovereignty and abolishing nationalism. It is about human freedom as necessarily a limited freedom.[3] And, it is about protecting our private lives within the four walls of a home where we are safe. For Arendt, limits and borders provide protection from dangers on all sides. And protection is a central concern in her thinking, understandably, perhaps, given that she fled for her life first from Nazi Germany and then from Nazi-occupied France.[4]

PLURALITY OF WORLDS AS THE LAW OF THE EARTH

Arendt opens *The Human Condition* with a description of the earth as "the very quintessence of the human condition" since it is most probably the only planet where we can "move and breathe without effort and artifice" (Arendt 1958a, 2). She defines the world, as opposed to the earth, as human artifice or what she calls our "man-made home erected on earth" (1958a, 134). In between earth and world, she introduces a third term, *nature*, which is associated with organic life "outside this artificial world," and through which "man remains related to all other living organisms" (1958a, 2). Although the earth provides the raw materials with which man builds his home, and nature supports life, both are indifferent to the plight of humans. Only our created world shields us from this indifference and protects us from the many dangers inherent in life on earth. And for Arendt, protection, security, and safety are the building blocks of home, which come through a sense of belonging. The earth itself, then, is not our home until we make it home; before then (if we can imagine a time before the world, which would be a time when we were animals and not yet human) while we lived on the earth, it was not home. And although this is where we belong in the sense that our bodies cannot live any place else, in Arendt's account, as animals, we do not yet have the sense of belonging inherent in feeling at home. Belonging, it seems, is a property of the world and not of the earth.[5] And world is distinctive of humankind. In the end, Arendt's sharp distinction between earth and world, particularly in terms of the association between earth and animal or body and world and human or mind, is problematic.

In an important sense, for Arendt, we do not share the earth, but only the world. Again, as animals on the earth, we share the planet with other species, but for Arendt, like Heidegger, animals are never properly *mit-sein*, and therefore *sharing*, properly speaking, is reserved for the human world, which by its nature is necessarily shared (e.g., Arendt 1958a, 176). Whereas in relation to the earth, homo sapiens are an animal species living among other animal species, in relation to the world we are human beings relating in meaningful ways to other human subjects. The world is always plural and shared; there is no private world in isolation. Arendt says, "No human life, not even the life of the hermit in nature's wilderness, is possible without *a world* which directly or indirectly testifies to the presence of other human beings. All human activities are conditioned by the fact that men live together" (1958a, 22). As an animal species, however, what she calls *animal laborans*, we are "worldless and herdlike" and therefore "incapable of building or *inhabiting* a public, worldly realm" (1958a, 160, my emphasis). As an animal species we *live* on the earth, and only as human beings do we *inhabit* a world. The world is what makes the earth a home for human beings. In other words, our life on earth—if we can imagine life apart from world—

may be necessary for sheer survival, but only when we create a world is it a meaningful life.

The world, says Arendt, "is what we have in common not only with those who live with us, but also with those who were here before and with those who will come after us" (1958a, 55). The world is described in terms of temporality, past, present and future. The earth, on the other hand, is the "limited space for the movement of men" and not identical with the world (1958a, 52). The earth is described in spatial terms as a limit to movement. Given that for Arendt stable products are the building blocks of the world, the world also takes up space. Even if earth and world exist in the same space and the same time, however, they are conceptually distinct, and the implications of this separation are profound. The earth is a limit condition on the world. It limits where we can move. It limits the natural resources with which we construct our world. It limits the conditions of life. Perhaps more important, as we will see, it limits the conceptual resources with which we build our world—and our worlds.

Arendt takes her analysis of the relation between earth, world, life, and home one step further when she maintains, "In order to be what the world is always meant to be, a home for men during their life on earth, the human artifice must be a place fit for action and speech" (1958a, 173). The earth, then, is the realm of *zoë*, or biological life, while the world is the realm of *bios*, or biographical life. And it is through narratives that we share a world.[6] In terms of the tripartite division of *The Human Condition*, we could say that labor or *animal laborans* gives us life, work or *homo faber* gives us humanity, but only deeds and speech or action gives us a home. Labor converts earth into nourishment for our bodies, work converts earth into natural resources and raw materials to build shelters and a stable world, but only action makes both earth and world a human home and a home for humanity. Action, says Arendt, "corresponds to the human condition of plurality, to the fact that men, not Man, *live* on the earth and *inhabit* the world" (1958a, 7, my emphasis). Action is possible because of plurality, because as human beings we are not only social beings but also diverse. The relationship becomes more complex when Arendt suggests that while politics is created by men and not by nature, men themselves are products of nature and of the earth: "Politics is based on the fact of human plurality," but "men are a human, earthly product, the product of human nature" (2005, 93). Given that we become properly human only by building a world in which we can act and speak—that is we become political—the phrase "human nature" is a an oxymoron for Arendt, which is why she talks of the human condition.

In order for humans to be at home on earth or in the world, we need the stable structures built by *homo faber*, along with the political rights built through collective action. Rights to equality and property are not gifts from nature or God, but rather gifts we give ourselves: "This voluntary guarantee of, and concession to, a claim of legal equality recognizes the plurality of men, who can thank themselves for their plurality" (Arendt 2005, 94). As political groups, we create the basis for equality, property, and all other rights that we extend to ourselves and to others. Although some have criticized Arendt for denigrating and devaluing the private social sphere in *The Human Condition*, it becomes clear in her later work that an essential feature of politics is to ensure the right to privacy to protect this realm. For Arendt this includes owning private property, to escape the public world, and have a place safe from prying eyes and from government intervention (1954, 186). Property for Arendt is not about dividing up the limited surface of the earth, but rather about having a place of one's own, a safe place to hide, a home. Certainly this need for home, for a safe haven, is something that we share with all of earth's creatures. The need for home is not unique to human beings. If we expand our horizons to include dens, nests, lairs, caves, burrows, digs, and holes, we realize that all animals need a safe haven to call home, whether or not their calls refer to anything as abstract as our notion of home (assuming that all human beings share a common notion of home, which, while we all may need or want one, does not mean that we share the same concept of home). Furthermore, if we extend our notion of home to include our environment and our natural habitat, then home eventually expands to include the earth itself and, in this case, home becomes the place we share with all living beings.

If Kant maintains that everyone has the right to property in the sense of the equal right to attain it, Arendt goes further and suggests that everyone has an equal right to private property in the sense of owning a place with four walls to call home. Everyone has a right to a place of her own. So too, Arendt insists on nation-states to ensure these rights. Although she is deeply critical of the homogenizing effects of nationalism and of defining the good in terms of patriotism—she criticizes Hitler for saying that what is good is good for the German people—she does not endorse a world constitution or global citizenship (1966, 300–302). On the contrary, she argues that cosmopolitanism risks the worst kind of world domination and tyranny and that only a plurality of nations can keep each other in check (1966). Politics is based on both the plurality of human individuals and the plurality of nation-states. Plurality operates as both a balance between individuals and states and a limit to individual and state sovereignty. "No man can be sovereign," says Arendt, "because not one man, but men, inhabit the earth" (1958a, 234; note that here Arendt says that we "inhabit the earth," whereas earlier we merely lived on the earth and inhabited a world). The same could be said of nation-states. No one state is sovereign over the entire earth because not one state but many exist on the planet. Only through the balance and limitation provided by plurality can we prevent totalitarianism and protect concrete rights. And, for Arendt, natality is what guarantees both plurality and diversity. Each one born is new and brings something unique to the world.

Arendt argues in *The Origins of Totalitarianism* that appeals to abstract human rights without recourse to concrete rights guaranteed by a state leave individuals unprotected: "The survivors of the extermination camps, the inmates of concentration and internment camps . . . could see . . . that the abstract nakedness of being nothing but human was their greatest danger . . . a man who is nothing but a man has lost the very qualities which make it possible for other people to treat him as a fellow-man" (1966, 300). Fellow men, it seems, are seen as such only when they share in the protections of a state, when they share in the sovereignty of a nation, balanced and limited by others. The human being qua human is not limited by the plurality and diversity of all people, but rather it is an abstraction, an extraction, of the most universal, generalizable, and therefore lowest common denominator, which for Arendt also makes it most like an animal in its mere membership in a species.[7] This is not necessarily to say that, stripped of its individuality or its nationality, the human is like an animal per se, but reduced to a mere member of a species, homo sapiens, or human being, its status becomes no more than any other animal. Stripped of its distinguishing features and of its nationality, it is no longer our "fellow."

Fellowship or recognition of our fellows, it seems, comes through political institutions that unite us with common goals or interests. When one is reduced to nothing more than a human being without those shared goals or interests, or, more accurately, shared rights and responsibilities circumscribed by civil codes, then one loses the fellowship of others. Insofar as fellowship is both social and political, it does not adhere to mere species being. Even while outlining the limitations and promise of the human condition, Arendt refuses to endorse a human nature that brings with it any natural rights. Rights are political and not natural. And yet human beings are naturally political beings, if not political animals, on Arendt's terms. For our shared animality is not what makes us political. Nor can it be the basis for fellowship. Rather, history, traditions, cultures, national identities, can be the basis for fellowship. Ironically, animals, too, share histories, traditions, and cultures, if not national identities. But, even the fellowship of animals cannot be reduced to their membership in a species. Following this Arendtian path, the growing attention to the many examples of cross-species altruism is noteworthy.

Continuing on this route, it is possible to think of our shared earthly home as a common goal and interest. The earth itself can become the basis for shared fellowship insofar as it is our home—our only home. And we all share it. The welfare of the earth, its health and sustainability, is in all of our interests. And, more than ever before, the environmental crisis and threat of climate change may bring humankind together as a species with a shared history and a shared future, which may depend upon coming together as a species, with concern for other species, to attend to our shared home, planet earth. Perhaps the cloud of climate change hanging over all human beings can bring us together beyond nation-states towards cosmopolitanism, not in the sense of a world government, but in the sense of a common cause, protecting and nurturing the earth. Perhaps Arendt's right to have rights can be grounded on the earth, not because of its limited surface, but rather because it is our home, our only home, the home that we share with all other living beings.

Arendt's insistence on both private property and nation-states is motivated by her concern with protection, security, and home. Both homelessness and statelessness leave people vulnerable, with no place to "hide" from the glaring intrusions of public life or the arbitrary dictates of the most powerful. Arendt insists that our safety depends on having places to hide (cf. Arendt 1958a, 71). The safety of the body depends on protecting it from the dangers of the inhospitable environment, including storms, heat, and freezing temperatures. But the safety of the body also depends on protecting it from the violence of other people, including crime and war. Arendt calls our bodies our most private property: "the body becomes indeed the quintessence of all property because it is the only thing one could not share even if one wanted to. Nothing, in fact, is less common and less communicable, and therefore more securely shielded against the visibility and audibility of the public realm, than what goes on within the confines of the body" (Arendt 1958a, 112). Just as the surface of the skin "hides and protects the inner organs that are [our] source of life," so too the walls of our house hide and protect our private lives (Arendt 1981, 25, 29). Arendt says, "These four walls, within which people's private family life is lived, constitute a shield against the world and specifically against the public aspect of the world. They enclose a secure place, without which *no living thing* can thrive" (1954, 186; my emphasis). Private property, or home, hides and protects the uniqueness of each human life. Every living thing needs a secure place, an enclosure that shields it. Every living being needs a home, whether it is a house or den, an apartment or nest, a tent or tree-stump. Indeed, it is telling that Arendt associates home with a place to hide, which we usually associate with animals. Like animals, we need a place to hide. Or all animals, including human animals, need a place to hide. And this place is a home. Home as a place to hide and a place to hide as home. We return to the issue of home in relation to all living creatures at the end of this chapter.

Both the private sphere of unique differences and the public political sphere of equality need to be protected.[8] In themselves the differences that makes us unique individuals, differences Arendt embraces as the essence of *natality*, threaten the political sphere, based, as it is, on equality. With her theory of natality, she brings together what makes each newborn radically unique—there never was or ever will be anyone else like that one—and the fact that each of us is born into a world of others, a world of plurality, a world we share.[9] She argues that because we are born we act, and, because we speak, plurality is the condition of our existence.[10] While natality is the hope for world renewal insofar as "the new beginning inherent in birth can make itself felt in the world only because the newcomer possesses the capacity of beginning something anew," this radical uniqueness seems in tension with plurality (Arendt 1958a, 9). Arendt puts it this way: "In man, otherness, which he shares with everything that is, and distinctiveness, which he shares with everything alive, becomes uniqueness, and human plurality is the *paradoxical plurality of unique beings*" (1958a, 176, my emphasis). Plurality rests on the fact of natality, through which our world is constantly renewed.[11] And yet the tension between the singularity of each one and the equality of all is productive insofar as plurality is possible only by virtue of natality, which is to say equality is possible only by virtue of singularity.

Equality born out of plurality cannot be a principle that levels individuality and makes each one the same. Rather, equality must confer on each—that is to say we must confer equality on each—the same rights not just in spite of our differences but rather because of them. "Plurality," says Arendt, "is the condition of human action because we are all the same, that is, human, in such a way that nobody is ever the same as anyone else who ever lived, lives, or will live" (1958a, 8). And, because we are plural, we are political.

Natality makes each individual unique and singular, but our social and political existence is based on our plurality.[12] In an important sense, much of Arendt's work is an attempt to navigate between the private realm of radical singularity and the public realm of radical equality. The private realm is not a protection for individualism, but rather for uniqueness and difference, for singularity. Arendt is clear that no individual exists in isolation from others. And the equality guaranteed by public state laws is not the equality of the lowest common denominator. Arendt distinguishes political equality from the leveling effects of statistics that turn people into populations and science that treats humans as nothing more than members of a common species.

Yet, in order to keep the private realm from becoming individualism or *each against all*, and in order to keep the public realm from reducing human beings to numbers, data, or specimens, we need a balance such that each limits the other. Again for Arendt the key to properly negotiating the various facets of human life is limits. Without the public sphere or the world, our earthly existence as unique beings has no meaning. And, without the private sphere wherein we attend to our existence, we are vulnerable to the harsh conditions of homelessness that threaten our very survival through the violence of nature. Statelessness, like homelessness, makes people vulnerable to violence: in this case, the violence of war, torture, and oppression. For Arendt, it goes without saying that animals are both homeless and stateless; they have neither private nor public lives. And yet it seems that, on Arendt's account, if animals have social codes and rituals that could be interpreted as political, then they too are world makers. And they too transform the given into the made, not only in terms of building or using tools but also in terms of forming a social structure with rules, or we could say with rights and responsibilities.

In *The Origins of Totalitarianism* Arendt describes the necessary tension between differences from birth that are given, on the one hand, and the equalizing of those differences that comes from being a citizen of a state and having rights that are made, on the other:

> The great danger arising from the existence of people forced to live outside the common world is that they are thrown back, in the midst of civilization, on their natural givenness, on their mere differentiation. They lack that tremendous equalizing of differences, which comes from being citizens of some commonwealth, and yet, since they are no longer allowed to partake in the human artifice, *they begin to belong to the human race in much the same ways as animals belong to a specific animal species.* The paradox involved in the loss of human rights is that such loss coincides with the instant when a person becomes a human being in general—without a profession, without a citizenship, without an opinion, without a deed by which to identify and specify himself—and different in general, representing nothing but his own absolutely unique individuality which, deprived of expression within and action upon a common world, loses all significance.
>
> (1966, 302, MY EMPHASIS)

In this passage Arendt indicates that what is at stake in political rights is not just protection from the violence of homelessness and statelessness but also the protection of a shared common world and access to the world of meaning. When she compares statelessness to animality, she is emphasizing the difference between biological life, or zoë, and biographical life, or bios, without which we are worldless. At stake, then, is not just homelessness or statelessness but also worldlessness. Worldlessness is the consequence of not having or respecting limits. In the case of homeless peoples, they are missing the limits to public surveillance provided by the rights to privacy and property. In the case of stateless peoples, they are missing the limits to dictatorial control of their bodies provided by law. At the extreme, in these cases they may be missing the limits to torture and genocide provided by international laws governing war.

TOTAL WAR AND THE PLURALITY OF WORLDS

Arendt argues that there are limits inherent even in war.[13] Following Kant, and thinking of war crimes, she maintains that nothing done in war should prevent future peace. But, thinking of nuclear war and genocide, she goes further than Kant and claims that nothing done in war should eliminate an entire world. What she calls "Total War," which aims to annihilate a race of people from the face of the earth, "oversteps" the limits of war, "limits that declared that the destruction brought about by brute force must always be only partial, affecting only certain portions of the world and taking only a certain number, however that number might be determined, of human lives, but never annihilating a whole nation or a whole people" (2005, 160). This genocidal war also oversteps the limits of politics to the point of annihilating the very possibility of politics because the "sheer existence of a nation and its people" is what is at stake (2005, 159). When this happens, then war is no longer a "means of politics" but rather "annihilates politics itself" (2005, 159). For Arendt, war must remain a means for nations and peoples to settle disputes through force that allows for the coexistence of the hostile parties and the continuation of their nations and their peoples. In other words, any war that puts an end to the possibility of war falls outside the bounds of the political.

For Arendt, the political is "based on plurality, diversity, and mutual limitations" (1968a, 81). Thus, if the aim of war is to annihilate human plurality and diversity, it has gone beyond the limits of what Arendt considers just war because it has gone beyond political aims. Moreover, genocidal war threatens the very possibility of politics. Given that an essential aspect of the human condition is that we are political beings who share a world(s) of our own making, to wage war on an entire subset of the human population is to deny one of the fundamental conditions of our own possibility as human beings, namely the plurality of our cohabitation. The "mortal sin," as she calls it, of this form of war is that it targets whole worlds, not only in the form of the products resulting from those worlds but also in terms of the culture, history, and traditions that make a meaningful world out of those products: "What perished in this case is not a world resulting from production, but one of action and speech created by human relationships. . . . This entire truly human world, which in a narrower sense forms the political realm, can indeed be destroyed by brute force" (2005, 161–162). This passage suggests that both *a* world and *the* world are destroyed by total war or genocide. While Arendt warns against the total annihilation of all human life and the destruction of the earth itself made possible by nuclear war, this is not what she means when she says that this entire truly human world can be destroyed. *A* world is destroyed when a culture and its people are destroyed. *The* world is destroyed when its witnessing structure based on the possibility of dialogic relations between diverse individuals and groups of people is destroyed. *Moreover, when a world in its entirety comes under attack, the world comes under attack, not just because every world contributes to the world but also and moreover because trying to eliminate any particular world from the world undermines the very structure of world itself.* The structure of world depends upon dialogic relations across diversity. Or, in Arendt's terms, the structure of the world depends upon speech and action, which are possible only because of human plurality.

Arendt's analysis of Eichmann and the Nazis can help clarify the relation between *a* world and *the* world. Arendt claims that because Eichmann "had been implicated and had played a central role in an enterprise whose open purpose was to eliminate forever certain 'races' from the surface of the earth, he had to be eliminated" (1992, 277).[14] It is important to note that in her remarks on Eichmann she suggests that a genocidal project may be the only one that justifies the use of the death penalty. As if speaking directly to Eichmann, Arendt concludes, "And just as you supported and carried out a policy of not wanting to share the earth with the Jewish people and the people of a number of other nations—as though you and your superiors had any right to determine who should and who should not inhabit the world—we find that no one, that is, no member of the human race, can be expected to want to share the earth with you. This is the reason, and the only reason, you must hang" (1992, 279).

In her extension of Arendt's analysis of Eichmann, Judith Butler argues that Eichmann's crime was thinking that he and his fellow Nazis could choose with whom to share the earth. Butler makes the case that human freedom is based on the fact that we cannot choose with whom to share the earth and that the unchosen quality of our cohabitation is essential to both the ethical and political demands of living together: "unwilled proximity and unchosen cohabitation are preconditions of our political existence," which brings with them "the obligation to live on the earth and in a polity that establishes modes of equality for a necessarily heterogeneous population" (Butler 2012b, 24). Even if we can choose with whom to associate, none of us have the right to choose with whom to inhabit the earth. Moreover, we have an obligation to affirm the existence of all others insofar as their very existence is a precondition for politics and we are political creatures. Prior to any contracts or deliberative volition or actions, we cohabit the planet with diverse others. And this cohabitation brings with it both political and ethical obligations to "actively preserve and affirm the unchosen character of inclusive and plural cohabitation" (125).

We might add that the reason we cannot choose with whom to cohabit the earth is because earth is our only habitat. As the only habitable planet for all earthlings, earth is not only our habitat and home but also that of every other living being. As earthlings, we have a singular bond to the earth. And, as such, we have ethical responsibilities to it and to its inhabitants. If we cannot choose with whom to cohabit the earth, this applies not only to fellow humans but also to all living beings, to all earthlings. Further complicating the ethical normativity of this cohabitation is that although we do not choose it, we have an obligation to choose it. We have an obligation to preserve and affirm it as actively choosing what we do not choose. That is to say, we have the paradoxical responsibility to choose the unchosen, which is the seat of ethics.

This is the ethical prescription inherent in Arendt's politics of plurality. If we are dependent upon others for the very possibility of having a world, then we are ethically obligated to sustain them, even if for our own sake. We might ask of Arendt and Butler whether we are required to actively preserve and affirm everyone with whom, and every world with which, we inhabit the earth. It is clear that Arendt does not think that we are required to affirm our cohabitation with Eichmann. Indeed, because he chooses not to share the earth or the world, no one should share with him. It seems, then, that at least in the case of those people or groups whose worldview includes the annihilation of another worldview, we are not obligated to preserve and affirm them. Yet we may still be required to preserve and affirm the unchosen character of the people of the earth in all of their diversity and plurality. In other words, we need to preserve and affirm the unchosen, diverse, and plural character of peoples and world that exist on the earth, even if we are not required by this ethical principle to preserve and affirm specific individuals or groups whose worldview includes denying or disavowing the unchosen, diverse, and plural character of peoples and worlds that coexist on the earth.

Moreover, just as equality is not given, but rather made by political institutions and guaranteed by civil laws, so too cohabitation must be affirmed as an ethical choice. This is to say that although the ethical ground of politics may be that we share the limited surface of the earth and that we do not choose with whom to share the earth, that ground becomes ethical only when we avow cohabitation and affirm it as an ethical decision. In other words, the facts of cohabitation alone are not enough to ensure ethical relationships or political equality or freedom. The facts of cohabitation must be continually interrogated, interpreted, and reaffirmed through questioning and analysis. In this way they become meaningful. For example, we might push the notion of cohabitation beyond the human sphere and talk about our cohabitation with animals or other living things. We could reinterpret cohabitation in a more expansive way that includes all the creatures of the earth. When we do so, our sense of ethical and political obligations shifts dramatically.

By returning to Arendt's texts in terms of the relation between earth and world, we might develop a more robust account of the ethical and political obligations inherent in the notion of plurality, namely *sharing the earth* and *cohabiting the world*.[15] In the passage from *Eichmann in Jerusalem* in which Arendt suggest that no one has the right to determine with whom to share the earth, she makes a subtle distinction between earth and world. She talks of *sharing* the earth and *inhabiting* the world, which is consistent with her discussion of earth and world in *The Human Condition* where she says that we *live* on the earth and *inhabit* the world. But Arendt also formulates the distinction in terms of desires and determinations, namely in terms of the Nazi's not *wanting* to share the earth and *determining* with whom they would inhabit the world. Obviously the two are related since she also speculates that because the Nazis did not want to share the earth with the Jews and certain others they acted as if they had the right to determine with whom to inhabit the world.

Returning to Arendt's articulation of the distinction between earth and world in *The Human Condition*, while we can, in a significant sense, determine our world, we cannot determine the earth. For Arendt, the earth is given whereas the world is made, or we could say that the earth corresponds to being whereas the world corresponds to meaning. There is a necessary gap between given and made, being and meaning, which should prevent totalizing ideologies such as Nazism. For when we mistake meaning for being we confuse world and earth, made and given. We mistake our worldview for the natural world and take our world to be the only possible world, the true and correct world opposed to all others. Naturalizing the norms and values of our culture or tradition can be used to justify eliminating anything that is "unnatural." Taken to the extreme, the collapse of meaning into being can lead to deadly justifications for killing others who do not share the same worldview, all in the name of eliminating what is unnatural or inferior by nature.

Approaching the question of the gap between being and meaning from the other side, so to speak, in *Natality and Finitude*, Anne O'Byrne argues that the gap between what she calls *being and knowing* or *nature and history* protects our freedom. She is concerned with how this gap prevents the human world from being a deterministic one governed simply by laws of nature. O'Byrne argues that although the modern worldview threatens to close the protective gap between being and knowing, Arendt maintains that it will never succeed because "the protective dividing line is never quite eliminated from our conceptual scheme and the human and natural worlds are never co-extensive" (O'Byrne 2010, 102; cf. Arendt 1958a, 324). O'Byrne invokes the notion of natality to explain why. In her brilliant analysis O'Byrne describes Arendt's natality as a "syncopated temporality" that opens a gap between nature and history through the future anterior tense of birth. In other words, since, in the phenomenological sense, no one is present at his or her own birth, each person's relation to his origin is subject to a retroactive temporality through which the event of his birth becomes meaningful. O'Byrne says that the fact that the moment of my birth is "irretrievably lost to my experience" makes the claim that we never have direct or complete access to our origins not only a philosophical claim but also "an indication of the very concreteness of the unreachable origin" (104). Furthermore, since this origin is structurally lost to any firsthand experience, it always necessarily comes to me through others. This means that I am for others before they are for me. "The origin from which I am removed is certainly mine, but it also belongs in an important sense to others. Our coming to be is therefore never a singular or solitary emerging into being; it is always, from the very start, a matter of plurality" (106). O'Byrne argues that the syncopated temporality inherent in natality is inherent in all of our experience insofar as every fact or event becomes meaningful to us only through interpretation. The time lag between the event and its meaning is also an ongoing one insofar as we can and do continue to reinterpret the past. In this sense, our origin is never past but always coming to be. O'Byrne concludes, "Action, not only in the political realm but also in the realm of nature, turns out to be belated in such a way that it holds open the gap between event and meaning. Action awaits its meaning in the same way that . . . birth only later comes to be my birth" (100). O'Byrne's account of syncopated temporality will become important for later discussions of the origins of both earth and world. For now, however, let's return to the dangers inherent in closing the gap between them.

In *The Human Condition* Arendt sets out some of the dangers of closing the gap between event, or being, and meaning, or knowing, which in the terms of this project we could call the gap between the earth and the world. We could describe closing the gap as a kind of confusion between being and meaning, given and made, or earth and world. This confusion is a dangerous category mistake that has practical and conceptual consequences that we should avoid. Perhaps we need further clarification on the distinction between worlds and World before we can understand the distinction between, and relationship of, earth and world. For, as we have seen, it is unclear whether Arendt thinks that we *inhabit* the earth at all or whether this mode of being is reserved for the habitable world that we create. And we still do not know what Arendt means by *the* world as opposed to *a* world. Indeed, to speak of *our* world, what does that mean? What makes a world mine or ours? How many people does it take to make a world or a people? And with whom do I share a/the world? How many worlds are there: one or many?

Arendt's analysis suggests that there is one *and* many. Although she repeatedly uses the phrase "a world" when discussing what human beings create, she also refers to "the world" as what human beings inhabit together. The possibility of different worlds may be most apparent when she says that, through the child, the lovers "insert a new world into the existing world" (1958a, 242). As we have seen, however, she insists that any world is shared; no one has a world of her own. Yet it is unclear how many it takes to constitute a world. How many people make up a world? Is the child a world unto itself, or do the lovers and the child make up their own world? Allowing that there is no plurality of one, how many does it take to make a plurality? This question is crucial in terms of Arendt's analysis of "total war" wherein she indicates that what is wrong with this type of war, as opposed to other forms, is that it attempts to eradicate an entire population along with its world. Presumably, what she would consider "just war," on the other hand, kills some members of a race or a group but does not set out to annihilate all of them. Recall that the "mortal sin" of total war is that it annihilates a world, its products, its people, and, more significantly, its history and culture, what Jacques Derrida might call its *archive* (cf. Derrida 1984). The question remains, how many can be killed without destroying a people or its world? When does a just war turn into total war? How can we draw the line between just war and total war?

In *The Promise of Politics* Arendt argues that although the Greeks totally destroyed Troy, in a sense Troy and the Trojans still live on in the tales of Homer, who keeps their memory alive. There Arendt maintains that Homer's song bears witness to both the victor and the vanquished alike—so in a sense the Trojan's do not perish. Rather they live in Homer's tales. And thereby "in a certain sense—that is, in the sense of poetical and historical recollection—he undoes that very annihilation" (2005, 163). In a nostalgic tone Arendt seems to lament that the heroic warriors and wars of old have been replaced by genocidal urges to wipe out entire civilizations and completely erase them from the face of the earth. The risk of massacring all members of a race or culture is one horrifying aspect of this type of war; another is the risk of erasing all memory of it. And it is the latter as much as the former that leads Arendt to reject genocide as beyond the limits of both war and politics. In other words, eradicating a population from the face of the earth is the terrible possibility that confronts us in total war and, arguably, in every war. But this horrifying prospect is distinct from destroying their world, which is to say destroying *a* world and its place in *the* world. A people may be destroyed or disappear, but their artifacts, literature, and/or pieces of their culture may remain. In a sense lost civilizations can be reclaimed through archeology and history and interpretations and reinterpretations of relics from the past. If a people can be destroyed, but their culture can live on, then we must again ask what constitutes a plurality for Arendt. Does politics require a plurality of people, or a plurality of cultures, or both? If some killing and war is justified, then does a plurality require merely representatives of every culture to satisfy the diversity requirement? If so, how many representatives ensure plurality? Here again, returning to the relation between earth and world may help think through and beyond these questions.

When Arendt insists that what total war threatens is the loss of an entire world, and thereby the loss of politics itself, she is not talking about sharing the earth. Rather, she is talking about inhabiting a world. When Arendt claims that the Nazis thought that they had the right to determine with whom to inhabit the world, what does she mean? Given her insistence on the diversity and plurality of cultures (along with individuals), we can imagine that the Nazis and Jews did not inhabit the same worlds, but rather inhabited radically different worlds. And yet these different worlds also coexist in *the* world. As we have seen, Arendt defines the world as "what we have in common not only with those who live with us, but also with those who were here before and with those who will come after us" (1958a, 55). *The* world, as opposed to *a* world, is what we share with all human beings, past, future, and present. A world, as opposed to the world, is what we share with some subset, or subsets, of all people on earth, past, present, and future. The world, then, is what we share with all people who ever lived and ever will live on earth (or elsewhere).

Arendt further specifies what she means by "having the world in common" when she describes the combined effect of multitudes of perspectives that together yield the real world: "For though the common world is the common meeting ground of all, those who are present have different locations in it, and the location of one can no more coincide with the location of another than the location of two objects. Being seen and being heard by others derive their significance from the fact that everyone sees and hears from a different position. . . . Only where things can be seen by many in a variety of aspects without changing their identity, so that those who are gathered around them know they see sameness in utter diversity, can worldly reality truly and reliably appear" (1958a, 57). Here Arendt specifies that the real, true reliable world is one that exists within various perspectives or, we might say, various worlds. *The* world is what these various worlds have in common. And each world is a "position" within the world. There could be some worlds, then, that do not have anything, or very little, in common with the real world, for example, the world of a person undergoing a psychotic episode. Note that Arendt does not say that the world is the sum of all other worlds or all perspectives, but only variety of perspectives. Again, we might ask how many perspectives her plurality criterion requires. How many perspectives must converge before we have a true, real, reliable world?

It is helpful to consider that Arendt's radical perspectivism not only comes out of a Nietzschean embrace of different perspectives but also a phenomenological approach to the world of objects as the combination of all possible perspectives on it. Whereas Husserl thought that through transcendental deduction and imaginative variation we could extrapolate from materials given to our senses to fill in whatever perspectives might be missing in order to get an idea of the thing as it appears to us, Arendt seems to insist on actual historical perspectives, which can be given only by different people through their various narratives or stories. In other words, the philosopher, cannot alone in her study, deduce all sides of any given object and thereby deduce how it appears to us. Yet, like Husserl, Arendt is concerned with how things appear to us. And, like Husserl, she emphasizes the role of imagination in constructing this world of appearances (see Arendt 1953, 391–392).[16]

As Anne O'Byrne explains in *Natality and Finitude*, Arendt is a *historical phenomenologist*. Her approach is one of taking up various perspectives from history and assessing how things appear in different historical epochs, along with presenting a phenomenological account of the meaning or essence of those epochs. Her historical approach to phenomenology operates to unsettle many aspects of the world that we take for granted or at least various parts of traditions that we may have thought we understood. In other words, Arendt uses history to show us different perspectives on the same thing, concept, or experience. Her appeals to history create an uncanny relation to what we think we know or understand. And yet, as we will see, it is through this uncanny and strange understanding that we may hope to create a home, not once and for all, but continually through questioning what we believe and what we take to be real, true, and reliable in our world.

There is a world only because there are perspectives. And there are perspectives only because there is plurality. Because human beings are diverse and relate to each other across differences, there is a world. "The world comes into being only if there are perspectives" (Arendt 2005, 175). And the more perspectives there are, the more world we have, not just in the sense of understanding the whole or the true world, or the real world, but also in the sense of enriching the meaning of the world. Arendt says, "the more peoples there are in the world who stand in some particular relationship with one another, the more world there is to form between them, and the larger and richer that world will be" (176). For Arendt, human beings are human by virtue of existing together in a world. And a world exists only through the plurality of human relationships. If this plurality or diversity disappears or is annihilated through war, then the world disappears. She couldn't be more forceful when she claims, "Human beings in the true sense of the term can exist only where there is a world, and there can be a world in the true sense of the term only where the plurality of the human race is more than a simple multiplication of a single species" (176). Without plurality, we are worldless.

We might push Arendt further at this point and argue that the human race is dependent upon the plurality of species and there is a human world only where there are interrelations between humans and other species. The plurality that is constitutive of the human condition extends beyond human diversity and into biodiversity. Even if, with Arendt, we separate world and earth by associating world with the human world of meaning and earth with our given physical limitations or animal bodies, still both the human world of meaning and our existence on the physical earth are dependent upon not just human diversity but also biodiversity. Our imaginations are fueled by our cohabitation with difference species, evidenced by our mythologies, literatures, and even our scientific research, inherently involved as they are with animals. Indeed, animal metaphors fill our language. And it is impossible to imagine what our world or worlds would be like without other animal species sharing our planet. Without our animal cohabitants, the life of the mind, as Arendt calls it, would be severely impoverished, and perhaps even impossible. In terms of our embodied existence, our very survival depends on other species, especially an entire universe of microscopic organisms with whom we share a symbiotic relationship not only on the planet but also within our very bodies. Without a much broader notion of plurality beyond mere human plurality, we cannot begin to understand what it means to inhabit either world(s) or earth. It is clear that Arendt's notion of plurality needs to expand to include other species, but the question remains of whether or not Arendt's demand for plurality as essential to the human condition applies to individuals, whether human or nonhuman. This problem is especially acute when we return to her insistence that some war can be justified so long as it is not a total war aimed at annihilating an entire race or people. And what of attempts by human to eradicate entire species, especially those seen as invasive, pests, or life-threatening such as bacteria and viruses?

Although in *The Human Condition* and throughout her writings she appeals to different historical periods and their worldviews, and only individual lives insofar as they are "heroic," Arendt's stance on total war suggests that she is also concerned with the survival of different perspectives in the mundane persons who hold them. On the other hand, in her analysis of Homer, she maintains that the Trojans—or, more accurately, their world—survive in the tales told about them. Given the passage from *The Human Condition* quoted earlier, we can conclude that only in a variety of different perspectives and what they have in common will we have the true and real world. Yet, taken to its limit, Arendt's definition of the world as encompassing past, present, and future entails that the world is in process and cannot be true and real until the end of humankind. In other words, the true or real world, like natality itself, is a matter of retroactive temporality or what Anne O'Byrne identifies with the future anterior tense, namely "it will have been." It "will have been" the real world only from the perspective of the sum of all perspectives, an impossible perspective at the end of time reminiscent of Kant's perpetual peace. For Arendt suggests, we only know the meaning of an event or a person's life after it has ended and so, too, we only know the meaning of humankind after the end of humankind: "The process of a single deed can quite literally endure throughout time until mankind itself has come to an end" (1958a, 233). We might conclude, then, that the real world is a concept but never a reality. Or that it is a regulative ideal in the Kantian sense. Or, perhaps, as Kant sometimes suggests, only aliens from another planet can see us as we truly are. Luckily we can skirt some of these thorny problems by recalling that the phenomenological approach teaches us to leave aside the metaphysical status of the real world and attend to how the real world appears to us. For, as we have seen, Arendt claims that it appears to us as what we share in common between various perspectives. Of course, the turn to phenomenology does not solve the problem of what counts as the real world. Moreover, it does not address the political question of who counts as *us*. When we consider how the world appears to us, does this us include all human beings, or even all earthlings, or just the ones most like ourselves?

Returning to the political question that drives much of her work, in this vein we might ask, what does Arendt mean when she suggests that genocide is an attack on *the* world and not just an attack on *a* world? This specifically political question not only falls more squarely into our present concerns than general reflections on the status of the world but also following this line of inquiry may shed light on the general problem of *the world* versus *worlds*. What genocide attacks, in addition to the physical bodies of members of certain groups or races of people, is the witnessing structure of the world itself. This means that by attacking *a* world in its entirety, we also attack *the* world, and not just because the world is the sum of many, or all, worlds. Rather, by attacking a world, we are also attacking the structure of the world and of world making. This means that we are attacking the ability of people or a people to construct meaningful narratives about their lives and thereby foreclosing the possibility of those lives becoming meaningful, which for Arendt also means becoming human. When she says, "This entire truly human world, which in a narrower sense forms the political realm, can indeed be destroyed by brute force," she is pointing out that the possibility of witnessing itself can be destroyed by brute force.

If we are by virtue of the plurality of others upon whom we depend, both proximally and structurally, then by cutting off the possibility of those relationships we undermine the ability of people or a people to formulate their own identities. If the structure of subjectivity, and thus the meaning of human life, is dependent upon witnessing or narrative, then by destroying the possibility of testimony as the story we tell about ourselves, we destroy the very possibility of being human beings. And, while this destruction is rarely ever complete, the risk is the annihilation not only a world but also the annihilation of the very possibility of the world or the possibility of any human worlds. The mortal sin, then, of total genocidal war is the attempt to destroy a world and thereby foreclose the witnessing structure of the world.

As I argued in *Witnessing* (2001), oppression and torture damage, even annihilate, what Dori Laub identifies as the *inner witness* necessary for the process of narrative or testimony to support itself (Felman and Laub 1992). The inner witness is produced and sustained by dialogic interaction with other people; dialogue with others makes dialogue with oneself possible. In *The Life of the Mind* Arendt emphasizes the importance of this inner dialogue to critical thinking. In order to think, speak, or act, we need the inner witness, which develops in our dialogic relations with others, what Arendt calls plurality. Address and response are possible because the interpersonal dialogue is interiorized. And, through this inner witness, we have a sense of ourselves as subjects. A sense of identity and subjectivity is possible by virtue of the structure of witnessing as the possibility of address and response has been set up in dialogic relations with others. Total war or genocide targets not only the narratives, stories, and archives of a people but also the very structure of witnessing that enables those people to speak to themselves about themselves, the structure of subjectivity and therefore the structure of the world.

In light of this explication of the witnessing structure of the world, what can we make of Arendt's indictment of Eichmann and the Third Reich? Recall that she says that they did not want to share the earth and acted as if they had the right to determine with whom they inhabited the world. They acted as if their perspectives determined the world and that they could eliminate from the world a certain perspective or perspectives that did not fit with their own. If, for Arendt, the world is a variety of perspectives, then they attempted to destroy not only *a* world or the world of the Jews, but also *the* world insofar as it includes the Jews and their world(s). Moreover, insofar as the Nazis tried to annihilate the Jews, they attacked the witnessing structure not only of the world of the Jews, who were literally and metaphorically rendered speechless, but also of the real and "entire truly human" world insofar as it is the result of the witnessing structure of human life. Resonant with Arendt's claims about the Nazis, in their analysis of survivors' testimonies Shoshana Felman and Dori Laub argue that the events of the concentration camps and mass murders ultimately became a Holocaust because they annihilated the possibility of witnesses and targeted the witnessing structure of subjectivity itself:

> The historical reality of the Holocaust became, thus, a reality which extinguished philosophically the very possibility of address, the possibly of appealing, or of turning to another. But when one cannot turn to a "you" one cannot say "thou" even to oneself. The Holocaust created in this way a world in which one could not bear witness to oneself. The Nazi system turned out therefore to be fool-proof, not only in the sense that it convinced its victims, the potential witnesses from the inside, that what was affirmed about their "otherness" and their inhumanity was correct and that their experiences were no longer communicable even to themselves, and therefore perhaps never took place. This loss of the capacity to be witness to oneself and thus to witness from the inside is perhaps the true meaning of annihilation, for when one's history is abolished, one's identity ceases to exist as well.
>
> (FELMAN AND LAUB 1992, 82, CF. 211)

Felman and Laub argue that the Holocaust attempted to annihilate the history of a people; and it did annihilate the history of individuals. The Nazis not only targeted the Jewish people and killed individuals but also annihilated the possibility of witnessing by destroying the address and response structure of subjectivity. In so doing they destroyed—or attempted to destroy—not only a world but also the world. Insofar as the world, the "entire truly human world," is constituted through interrelationality born of response-ability or the ability to respond, and insofar as the Holocaust destroyed the possibility of bearing witness even to oneself, it also attacked the very possibility of world. Thus, when Felman and Laub say that the

Holocaust created a world in which one could not bear witness to oneself, they are talking about a worldless world, a world in which the very possibility of world is called into question, a world turned against itself, in what Derrida would call the logic of autoimmunity: a world bent on destroying itself. As we will see in the next chapter, insofar as witnessing is based on address and response, it extends beyond human narrative and stories and encompasses the responsiveness of all earthlings along with various modes of address well beyond human understanding or human perception. But for now, let's return to Arendt's claim that the Nazis did not want to share the earth and the suggestion that the unchosen nature of earthly cohabitation is the ground of both ethical and political freedom. One way to approach these questions is from the perspective of "the given" as the unchosen in Arendt's thought. The unchosen and unchangeable given reminds us of our own limitations. In *The Origins of Totalitarianism* Arendt describes the relationship between the given and politics: "The dark background of mere givenness, the background formed by our unchangeable and unique nature, breaks into the political scene as the alien which in its all too obvious difference reminds us of the limitations of human activity—which are identical with the limitations of human equality." (1966, 301). For our purposes, we could say that the earth, and what is associated with it, is what is given, while the world, and what is associated with it, is what is made. What is given operates as a limit condition for what is or can be made. Certainly, Arendt thinks that we cannot change what is given. Indeed, she suggests that we need to find ways to accept—even to be grateful for—what is given.[17]

The given, like the earth, is a limit to the world and therefore to politics.[18] And only by respecting its borders, namely what is given and cannot be changed, are we free.[19] Freedom is not the freedom to do anything that we want. Rather, freedom is freedom only because it is bounded by what we cannot control and what we do not and cannot master. And even while our relationship to the given may change and evolve, and even while we have a responsibility to interpret the given as it affects our lives, we do not have the power to control or master it, perhaps not even to understand and know it. Even if we cannot distinguish the given from the made, even as we challenge this distinction and the limits between the two, we must acknowledge our own limitations insofar as we are not the masters of either. In Arendt's terms, living with the limitation of the given is an essential part of the human condition. The earth is the ultimate given in that we are all earthly embodied creatures who depend entirely on the earth for our existence and for our freedom. In *Between Past and Future* Arendt says, "Conceptually, we may call truth what we cannot change; metaphorically, it is the ground on which we stand, the sky that stretches above us" (1954, 263–264).

Given our analysis so far, we could say that what we cannot change is not just metaphorically the ground on which we stand and the sky that stretches above us but also literally the earth and the sky, or more precisely, our dependence on them. The problem comes when we try to change the earth and make it in our own image like we make the world and worlds. When we do so, we do not respect the limits of our own will, that is to say, the limitations of the human condition. The Nazis did not respect the limits of the human condition. In their arrogance, they believed that they could master the world and the earth. Rather than limit their own will, they exercised their will as if they could decide with whom to inhabit the earth; as if their world were the only world, which they would prove by annihilating all others. As Arendt makes clear, politics necessitates limits and respect for borders, conceptual, national, and individual. By imagining themselves limitless, the Nazis not only took the lives of hundreds of thousands of individuals and attempted to annihilate an entire people, but also threatened to destroy the very possibility of politics and, along with it, what makes the world an "entire truly human world." They broke the fundamental "law of the earth," namely, plurality.

EARTH AND WORLD ALIENATION

The lack of limits leads to both what Arendt calls *earth alienation* and what she calls *world alienation*.[20] Taking a closer look at these two concepts and the differences between them again shows us how Arendt's is a philosophy of limits and why we need to think earth and world together. In *The Human Condition* Arendt concludes, "while world alienation determined the course and development of modern society, earth alienation became and has remained the hallmark of modern science" (1958a, 264). She describes world alienation as the result of confusing private and public, social and political, and labor and work. Although these spheres—or we might even say worlds—cannot be easily separated in our experience, Arendt insists that conceptually they must be limited and contained in order to maintain distinctions between them. Although they may exist in the same space and time, they are not the same. And there are dangers associated with confusing them, which is why Arendt identifies the purpose of her historical analysis as tracing modern world alienation's "twofold flight from the earth into the universe and from the world into the self" (1958a, 6). Arendt associates the flight from the world into the self with modern philosophy and its turn inward, on the one hand, and the turn to the body as the focus of public life, on the other. Specifically, Arendt argues that the reductive focus on the body as a biological specimen over all other aspects of life reduces the meaning of life to mere survival, evidenced perhaps in contemporary culture by the discourse of "survivors," as if life can be defined in terms of mere survival. Indeed, for Arendt to define life in terms of survival is to evacuate it of its meaning and render us worldless. To have a world is to have meaning and to have meaning is to have a world. Worlds—plural—are the meaningful frameworks, literal and metaphorical, through which we interpret *the* world.

Arendt describes world alienation as the loss of a stable world of manmade meaning and therefore the loss of a common world that we all share. This loss of a shared world, or worlds, comes about through confusion between life and world or what she calls labor and work. Life and labor deal with the nourishment and survival of the body based on consumption of perishable goods; while the world and work deal with durable products whose stability provides the space in which we can speak and act. Ideally, products of labor are thoroughly consumed and perishable such that they are part of an ecosystem in which all waste produced by this consumption can be reabsorbed. The products of work, on the other hand, can be used, even used up, but never completely consumed or destroyed.[21] When we confuse the two, we seek abundance through accumulation and consumption, on the one hand, and turn durable goods into consumer goods, on the other.[22] It is telling that today what Arendt identified as a consumable good, bread, in some cases lasts longer than what she identified as a durable good, furniture. The preservatives in our food can make it last for years, but compromise the health of our bodies and the planet. While durable goods such as electronics, by design, become obsolete nearly as soon as they are released into the market. And both contribute to pollution; the first in terms of chemical waste, the second in terms of material waste piling up in landfills.[23]

Another Arendtian limit is that between the family, or the social, and the political. Again, although in our experience they are lived together, like labor and work, or perishable goods and durable goods, conceptually we must keep them separate. The state should not be considered one big family with the head of state as the patriarchal authority. Otherwise we risk totalitarianism. Moreover, when the nation is seen as one big family, the result is nationalism and patriotism that excludes some. Confusing these realms can be dangerous for those stateless peoples who have lost the protections of the state (see Arendt 1958a, 256). In a sense, Arendt's distinctions are based on her historical phenomenological account of the essence of different activities and historical eras. And to confuse these activities can have dangerous consequences for both earth and world.

What Arendt calls earth alienation is caused by the scientific worldview symbolized by Einstein's "observer who is poised freely in space" (195, 273). The view from the universe gives us the illusion that we are not earthbound creatures, but universal citizens who can leave earth. Arendt warns that we split atoms and take elements from outside earth into earth at its own peril: "And even at the risk of endangering the natural life process we expose the earth to universal, cosmic forces alien to nature's household" (1958a, 262). Earth alienation, like world alienation, comes through confusion between labor and work. With science, we think that the given world is man-made or can become man-made. We think that we create the earth and its raw materials. We mistakenly think that we are the masters of the universe and we do not acknowledge that we neither know the consequences of our actions, which are unpredictable, nor that we cannot undo what we have done. In a sense, earth alienation is the result of scientific hubris and the disavowal of the limits of the human condition.

Specifically, Einstein's free-floating observer is a denial of the fact that human beings can live only on earth. Moreover, the idea that we might someday find another planetary home is a denial of the fact that the human life span, at least today, is not long enough, even traveling at the speed of light, to escape our galaxy (Arendt 1954, 276). Furthermore, Arendt argues that even if we were to reach another standpoint from which to view Earth, whether it is literally another planet or Einstein's metaphorical observer, the scientific drive to master the entire universe will not stop there but demand ever further standpoints. The conquest of space as "the search for a point outside earth from which it would be possible to move, to unhinge, as it were, the planet itself" is never ending. "In other words," says Arendt, "man can only get lost in the immensity of the universe, for the only true Archimedean point would be the absolute void behind the universe" (Arendt 1954, 278). But, this view from nowhere is not only physically and intellectually impossible but also dangerous in that it presumes that man can take a God's eye view of the universe. This hubris does not respect the limits of the human condition. Furthermore, it risks not respecting the limits of the earth and of our world in relation to it.

THE "CONQUEST" OF THE UNIVERSE WITHIN

Our "conquest" of space, as Arendt calls it, has lead to attempts to harness extraterrestrial energies and bring them back to Earth, presumably where they do not belong. Unlike the authors of the *Time* article from 1969 honoring the Apollo 8 astronauts, for Arendt "conquering space" does not have a positive valence. To the contrary, Arendt warns of its dangers: "we release energy processes that ordinarily go on only in the sun, or attempt to initiate in a test tube the processes of cosmic evolution, or build machines for the production and control of energies unknown in the household of earthly nature" (Arendt 1954, 279). Today scientists are more focused on the universe within the human body than the universe beyond earth. Still, genetic engineering and DNA manipulations exhibit the same conquering tendencies diagnosed by Arendt. Man fancies himself master of his own creation such that he can create new human beings through genetic science that are "better" than what nature makes. Bioethicist John Harris makes this type of argument (Harris 2010). Parents are becoming patents in a world where DNA can be patented (cf. Oliver 2012).

For example, in 2013, the Supreme Court heard a case that may decide the future of DNA patents. The DNA in question involves two genes that significantly increase the risk of breast and ovarian cancers. The Myriad Corporation has patented the genes and a set of diagnostic tests to detect their presence (Kevles 2013). Patent law, which dates back to debates between Thomas Jefferson and James Madison, stipulates that patents can be obtained for "any new and useful art" or "process" (Kevles 2013). The requirement of newness and innovation excludes "products made by nature, which were held to belong to everyone, [and] were not to be removed from common possession" (Kevles 2013). Natural elements, creatures of earth, and natural laws are excluded from patents. For example, "natural elements taken from the earth, even if they had to be chemically isolated from other substances, [do] not constitute patentable subject matter under Section 101, if only because they were not new" (Kevles 2013). At issue in the case of the two genes patented by Myriad is whether the DNA sequences count as natural or man-made and, new to patent law, whether patent monopolies pose dangers to the rights of many versus the rights of few. The court ruled that DNA cannot be patented. From an Arendtian perspective, the attempt to patent DNA seems like a confusion of the given and the made and comes up against rights to private property. Indeed, patent law itself seems to push up against property law, which, as we have seen in terms of our analysis of Kant, is based on original acquisition of portions of the surface of the earth. If the earth itself can be delimited into private property, we might ask, then why can't DNA?[24]

Arendt, the philosopher of limits, insists that whatever new property we claim, whether outer space or the inner space of genetic code, this property will be limited by the facts that the earth is our home and we are mortal.[25] For, whatever our scientific achievements, these two facts do not change. We are limited mortal creatures of the earth. Even if the Mars One project succeeds in its plans to send earthlings on a one-way trip to Mars in 2023, everything they experience there will be judged and understood in terms of their experience on Earth. Even if we set up colonies on Mars, we are still creatures of the Earth, at least for the foreseeable future. For now, earth is our only home. And, both earth alienation and world alienation leave us homeless. When we are alienated from the world, we disavow or deny that the meaning of life comes through our relationships to others. Our world becomes meaningless, a worldless desert that is barren and isolating. Worldlessness is a loss of meaning between human beings. When we are alienated from the earth, we disavow or deny that we survive and thrive as a result of our unique and irreplaceable bond to the earth and its creatures. When this happens, we are truly homeless insofar as we no longer feel at home on the earth. Both forms of alienation are maladies of perspective that result from a loss of limits. When we refuse to acknowledge and respect our limitations, then we risk the homelessness that results from both earth and world alienation. When we mistaken believe that we can master, control, and manage earth and world, we lose our home in both. As Arendt suggests, only from some perspective outside of earth or outside of world can we imagine grasping them as we might an object and subjecting them to our technologies in order to master them. And, while we do this all the time, the risk is that we will forget, or deny, that earth and world are more than objects for our control, manipulation, or management—they are also our home. In a sense, the earth nourishes our bodies, while the world nourishes our souls. By denying our limitations and continually trying to overcome them, we deny the human condition, bounded as it is by both earth and world. Returning home, then, would be a process of acknowledging our limitations and exercising self-restraint. This is particularly apt in terms of our relationship with the earth and the environment. Finding our way home is what Arendt calls the "strange enterprise" of imagination through which we can change perspectives and change worlds.[26]

UNDERSTANDING HEART

In *Understanding and Politics* Arendt gives imagination the central function of putting things in their "proper perspective" by negotiating between the immediacy of experience and the distant abstraction of knowledge, negotiating between the living body and that free-floating point in space. She describes this "proper perspective" as neither too close nor too far away. Imagination allows us to "see" what is close to us as if it were at a distance, but it also allows us to see what is far away as if "it were our own affair." She says, "This 'distancing' of some things and bridging the abysses to others is part of the dialogue of understanding for whose purposes direct experience established too close a contact and mere knowledge erects artificial barriers " (1953, 392). Suspicious of her notion of a "proper" perspective and our ability to understand completely or once and for all, I agree that imagination is crucial to taking up varying perspectives that allow us to continue to question and interpret our experience and our relationships with others and with the world.

Arendt places imagination between feeling and reflection as the dynamic flowing blood pulsing through what she calls an "understanding heart." She identifies the acceptance of the given as an affirmation of what is, an affirmation that can lead to gratitude, provided we have an "understanding heart" (1953, 391). Only through the "strange enterprise" of understanding, dependent as it is on imagination, can we "come to terms with what irrevocably happened and be reconciled with what unavoidably exists . . . only an 'understanding heart,' and not mere reflection nor mere feeling, makes it bearable for us to live with other people, strangers forever, in the same world, and makes it possible for them to bear with us" (391). In order to live amongst others and embrace our plurality, we need an understanding heart. We must create our world and worlds with this understanding heart such that we accept, even embrace, the given, most especially the diversity of human existence and, I would add, the biodiversity of our shared planet. Furthermore, adding the force of an ethical imperative, we could say that we have an obligation to create worlds—and thereby the world—in such a way that we embrace what is given, most especially our unchosen cohabitation with other people and other creatures of the earth. Ethics starts when we embrace our limitations and humble ourselves before the magnificence of earth and its inhabitants. Only when we attend to the earth and other earthlings with whom we share our planet in ways that open up rather than close off the possibility of response, can we begin to act responsibly. Reconciliation to the limits of the human condition, is not, then, giving up or throwing up one's hands in the face of what cannot be changed. Rather, it is embracing and affirming it, but not in an unreflective or naive way. Moreover, it is always a matter of interpretation. The given is never merely or purely given. Facts always require interpretation, and this is where imagination and understanding interplay with perception and the basis of perception in the limitations of our bodies. Certainly, our relationships with other people—and with other creatures of the earth—is not merely a matter of perception alone. If we associate what is given with the earth and what is made with the world, this means that for us the earth always requires a world in terms of which to view it. There is no view from nowhere, no disembodied astronaut viewing the earth from an abstracted point in space. Rather, there are only specific locations in space and time from which we view the earth, and everything on it, from our place in the world. This is why Arendt's separation of earth from world is a *conceptual* distinction that functions in a particular worldview she develops out of her historical phenomenology. So, when she proclaims, "plurality is the law of the earth," that pronouncement is part of a world in which the earth itself is considered both the ground and limit to politics and, we could add, to ethics (1981, 19).

For Arendt, finding our home may involve an endless journey visiting different worldviews in order to find ourselves situated—and, more to the point, *situating* ourselves—amongst them. It may involve wandering to the ends of the earth, metaphorically if not also literally. Although the earth operates as a limit concept in Arendt's philosophy, it is also a planet and the only one upon which we can live and build our home/s. It is noteworthy in this context that the word *planet* is from the Greek *planetes*, meaning "wanderer."[27] What might it mean to think that a wanderer is our home? Or that we have a wandering home? Or even that we are wanderers on this wandering planet that is our home? Perhaps there is a necessary element of the alien or foreign in every home; as Julia Kristeva says, we are strangers to ourselves. Moreover, we could imagine that being too much at home may lead to inaction or a contentedness that runs contrary to both ethics and politics in terms of the struggle for social justice. After all, Arendt imagines that "the danger lies in becoming true inhabitants of the desert and feeling at home in it," when this desert is "the growth of modern wordlessness, the withering away for everything between us" (2005, 201). She is clear that we should not become reconciled to the desert; rather we must protect the "life-giving oases" "that let us live in the desert without becoming reconciled to it" (203). These life-giving oases are love and friendship, which allow us to endure the desert of public and political life. And, in a way, that love and friendship carries over into our world-building activity as what Arendt calls our *amor mundi* (love of world).[28] She concludes, "In the last analysis, the human world is always the product of man's *amor mundi*, a human artifice whose potential immortality is always subject to the mortality of those who build it and the natality of those who come to live in it" (203).[29]

But, we might ask, what does Arendt mean by *amor mundi*? And what kind of love is love of the world? Arendt is clear that romantic or sentimental love is a private affair and not part of the political world. Yet, she also names love, along with friendship, as one of the oases in the desert of worldlessness that can restore the meaning of the world. Furthermore, she acknowledges the importance of intimate relations to sustain one's ability to enter into the public world and perform in the world of politics. What then is the relationship between love and politics if what binds us to the public world is love, *amor mundi*? Perhaps love of world is the primary political virtue. We might answer that love of the world is necessary in order that politics becomes ethical. And that love of the world is the foundation for Arendt's famous claim for "the right to have rights," which is to say, ultimately *amor mundi* is the force behind universal or transnational norms.[30] For "the right to have rights" goes beyond any legal authority, national or transnational, and takes us into the realm of ethical norms grounded in our very belonging together on the earth and in the world.[31] This is the ethics based on our shared home, our cohabitation on the earth, and the unchosen nature of that cohabitation that compels us to choose it—even to love it. Love of world is the embrace of this plurality that makes world and worlds possible.

Insofar as Arendt emphasizes the importance of forgiveness, along with promises, to the political bond, her invocation of love, at least implicitly, resonates with the notion of *agape* as forgiveness. And yet her appeal to Augustine and his proclamation "I want you to be" suggests a form of love beyond forgiveness and moving toward acceptance. Yet even acceptance of the existence of the other or stranger is not enough to make the claim "I *want* you to be." Otherwise, Augustine could say, "I accept that you are." Embracing the existence of the other and our coexistence or cohabitation with others, including other creatures, takes us beyond either forgiveness or acceptance and toward love. To be political and, in our framework, also ethical, this love is more than romantic or sentimental love. Rather, love of those with whom we cohabit the earth—along with love of other living beings and perhaps even nonliving things—is an ethical and political choice. The paradoxical situation in which we find ourselves, which is essential to the human condition, is that although we do not choose with whom to inhabit the earth ethical and political bonds require making that choice. Choosing to love. This is to say, ethical relations to others require the Augustinean proclamation "I want you to be," and political bonds require extending this proclamation to whole groups of people, perhaps including nonhuman animals. Imagine what it would be mean to say to all animals or all earthlings, "I want you to be." This would be an ethics of affirmation of all and each. In this regard, Arendt's extension of the Augustinean proclamation becomes the basis for an ethics of coexistence or cohabitation that grounds all political claims. Arendt's notion of *amor mundi* signals not only a love of one's own world but also, and moreover, a love of the world of others. For Arendt, worldliness is always a matter of difference, diversity, and coexistence or cohabitation with others unlike myself. *Amor mundi* is an embrace of the diversity of the world and of worlds.[32] It is also an acknowledgment of our deep dependence on the plurality of the earth.

In this regard, we may think of love in terms of eros as the life force that compels us to bond with others.[33] Eros takes us out of our selves and toward others, and through connections with others, human and nonhuman, we not only survive but also thrive as individuals and as species. For Plato, Eros is a form of love as passion that gives rise to creativity and the highest forms of contemplation. The tensions inherent in Eros move us toward something beyond ourselves. For Arendt, it is our relationships with other people that give rise to creativity, contemplation, and move us toward the political bonds through which we peacefully, more or less, cohabit. This creativity born from love provides an oasis of meaning in the desert of meaninglessness. That meaninglessness ensues from denying relationality, plurality, and the dynamic nature of the political world. Ultimately, for Arendt natality is the concept that signifies the plurality and diversity of the world as enabling both creativity and contemplation. The unpredictability of birth, which even in the most controlled and usual of circumstances results in the birth of a unique individual who is a stranger to her parents, becomes the symbol for the uncanny strangeness of the world. We might compare this aspect of natality, namely the necessity to choose the unchosen, to Emmanuel Lévinas's notion of "paternal election," which is the choice of this particular unchosen, even unbidden, child (1969; cf. Oliver 2011). Can we expand this election of one particular child to every human being and beyond? Can we elect each and every earthling because it is born, hatched, spawned, and therefore unique? Certainly, Arendt would never go so far since she reserves birth and uniqueness for human beings alone (e.g. Arendt 1958a, 176). Yet what is more uncanny than an encounter with another species?

Through friendship and love we not only come to terms with this uncanny strangeness but also learn to embrace it, which is possible only when we give up the fantasy of being able to control or master it. Interpreting Augustine's love as affirmation of existence and cohabitation, Arendt says, "This mere existence, that is, all that which is mysteriously given us by birth and which includes the shape of our bodies and the talents of our minds, can be adequately dealt with only by the unpredictable hazards of friendship and sympathy, or by the great and incalculable grace of love, which says with Augustine '*Volo ut sis* (I want you to be),' without being able to give any particular reason for such supreme and unsurpassable affirmation" (1966, 301). The ethical affirmation of each that grounds the political bond comes from the heart and not reason alone. Through natality, which Arendt calls a *miracle* that saves the world, this affirmation renews the bonds of the political world.[34] Although it is not always the case in practice, in principle, through the birth of each unique being, we come to accept, even love, the newcomer who is at first always a stranger. Natality renews the political world by injecting plurality with diversity since every individual is unique. And plurality, says Arendt, "is the law of the earth" (1981, 19).

If plurality is the law of the earth, and politics is based on this plurality and diversity as the affirmation of each one, what are the implications of shrinking biodiversity for politics? What happens when climate change leads to the spread of a literal desert along with Arendt's metaphorical one? Indeed, what do these metaphors of an inhospitable world tell us about our need for protection from the violence and harshness of both nature and culture, both earth and world? Could the earth itself become the "you" in Augustine's "I want you to be"? To answer these questions, we must extend Arendt's philosophy of limits in order to develop a sustainable ethics of conservation through which we impose limits on ourselves for the sake of both earth and world in an attempt to actively engage in what Heidegger calls the "letting be" of beings. In "The Crisis of Education," Arendt claims, "Education is the point at which we decide whether we love the world enough to assume responsibility for it and by the same token save it from that ruin which, except for renewal, except for the coming of the new and young, would be inevitable" (1954b, 196). We might ask, what would it mean to love not only the world but also the earth enough to assume responsibility for it? Can we imagine a world in which the earth matters enough that we take responsibility for it? Perhaps only a cosmopolitanism of citizens of the earth can assume such responsibility, a responsibility that Arendt finds daunting. And yet the environmental crisis may require a world of this magnitude. Can we imagine a cosmopolitanism that does not deny differences or level diversity, but rather one through which we embrace the singularity—the natality—of each living being? Moving from Kant to Arendt, and through Heidegger and Derrida, hopefully, this is the path that we are following.

COSMOPOLITANISM AND HUMAN SOLIDARITY

In her analysis of Karl Jaspers's cosmopolitanism, even while she rejects the notion of world citizens, Arendt is sympathetic to the idea of human solidarity.[35] And even while she is critical of the unity of mankind, she supports the inherent connections between men: "A philosophy of mankind is distinguished from a philosophy of man by its insistence on the fact that not Man, talking to himself in the dialogue of solitude, but men talking and communicating with each other inhabit the earth" (Arendt 1968a, 90). This is why she says, "plurality is the law of the earth" (1981, 19). In the end she comes close to endorsing Jasper's historical and political concept of mankind and world citizenship, which she contrasts with Kant's ahistorical and Hegel's apolitical views (Arendt 1968a). She imagines Jasperian unity brought about by "technical mastery over the earth," symbolized by the "possibility that atomic weapons used by one country according to the political wisdom of a few might ultimately come to be the end of all human life on earth" (83). While she maintains her belief in the protective checks and balances on any nation-state by all the others, she entertains the idea that some sort of "world-wide federated structure," and at least international law, is necessary to address the crisis of the possibility of nuclear destruction, which threatens all of humankind and the earth itself. North Korea and Iran notwithstanding, we may wonder whether the environmental crisis and climate change have replaced the destructive threat of total annihilation by atomic bombs and that, even more than the nuclear threat, the environmental threat requires some notion of world citizenship, not only with its rights but also with its responsibilities.

Discussing Jasper's call for a politics of human solidarity, with some approval Arendt says, "Just as, according to Kant, nothing should ever happen in war which would make a future peace and reconciliation impossible, so nothing, according to the implications of Jasper's philosophy, should happen today in politics which would be contrary to the actually existing solidarity of mankind" (Arendt 1968a, 93). This view of human solidarity, however, entails that *no war* is justifiable "not only because the possibility of an atomic war may endanger the existence of all mankind, but because each war, no matter how limited in the use of means and in territory, immediately and directly affects all mankind" (Arendt 1968a, 93). Arendt worries about the complete abolition of war as a political means of maintaining a balance between nation-states. She sees the option of war as a protection against totalitarianism. When war is about annihilation rather than about politics, however, it exceeds its own limits and the limits of politics in an absolute way that she argues makes it impermissible. Yet, insofar as all war "affects all mankind," it is not permissible. This is to say, insofar as war goes beyond a political means of settling a dispute between nation-states, it oversteps its bounds. In our increasingly globalized world, the affects of war in one part of the world are felt in various ways in others. In addition, as recent United States history has shown, the rhetoric of fighting totalitarianism or overthrowing dictatorship can become a call to war for all sorts of reasons, many of which have nothing to do with liberating people from totalitarianism.

Moreover, taking this thinking further, insofar as every war threatens the structure of the world, all war is impermissible. All war threatens the existence of politics. As we have seen, Arendt argues against "total war" on the basis of its attack on entire worlds and peoples and its attempt to annihilate histories and traditions along with killing enemies. But it is impossible to draw the line between just war and total war on these terms. For how many or how few can be killed before we risk destroying "a people" or a culture? How many does it take to make up a plurality? How many does it take to make a world? Furthermore, insofar as war threatens not just the lives of others but also their ability to response, insofar as it destroys their inner witness and therefore their ability to bear witness for themselves, war threatens the very structure that makes living in a world possible. All war, then, threatens the "entire truly human world." Taking Arendt's discussion to its limit, we could say that there is no war that can respect that plurality is the law of the earth.

While Arendt comes close to endorsing Jaspers's view that human solidarity should guide politics in order that no state act in such a way that human solidarity becomes impossible, she stops short of arguing for the abolition of all war. Yet, if we cannot define plurality in terms of absolute numbers, then it is difficult to maintain both Arendt's insistence on war as a political option at the same time as her adherence to a principle of plurality based on cohabitation. In other words, if, as Arendt insists, war that seeks to annihilate an entire people or race and destroy its culture is a hard limit for politics because it attacks an entire world, then again we must ask, how many does it take to make a world? Arendt discusses plurality and diversity, but does not stipulate how many different people are necessary to form a political union and make up a people or a world. Indeed, as we move through Heidegger to Derrida, the question whether each singular individual living being constitutes not only *a* world but also *the* world is definitive. Even for Arendt, however, the self-definition of groups is not dependent upon their numbers. Rather, worlds are formed through relationships between people (including nonhumans and the environment). Don't we have to consider, then, the ways in which worlds may be threatened and destroyed by any war?

If, as Arendt suggests, we must make war in ways that always allow not only for future peace (as Kant might say), but also for future wars, then we must make everything we make in ways that allow for future production. Certainly, war should not be the only renewable resource. War affects not just human beings but many other beings on the planet. And the war machines of the military industrial complexes across the globe impact not only our natural habitat but also that of other earthlings. Moreover, if we adopt human solidarity as our guiding political principle, then it is not only war that threatens worlds. Today, more than the threat of nuclear war, the threat of environmental crisis threatens human worlds, along with nonhuman worlds, whatever they may be. Indeed, the notion of human solidarity born from seeing the first images of earth from space, images that inspired environmentalism, can perhaps become more than the cold war rhetoric justifying one nation's technological advances over another. Rather, the realization that not only do we share the earth with all other humans, and all other living beings, but also that it is our only home expands the notion of human solidarity.

Human solidarity becomes grounded on the earth to which we belong. The connection between peoples and nations across this planet means that actions by one can affect distant others in ways that threaten human solidarity. Given that all humans inhabit the earth, and that, in spite of our differences, we share a special bond to this planet and none other, the possibility for human solidarity is bound up with the ways in which we cohabit the earth. And, yet, the realization that this is not just our home, a human home, but also home to every other living creature enlarges human solidarity beyond the human. Now we must imagine the solidarity of earthly creatures, earthling solidarity. Earthling solidarity requires imagination associated with the understanding heart, this is to say, imagination that reaches beyond the self and toward others, now not only human others but also nonhuman others, by first acknowledging that we live on this planet together. In this regard, loving the world enough to take responsibility for it entails loving the earth enough to take responsibility for it.

Arendt argues that only through this understanding heart or loving imagination can we take our bearing in the world. She calls this loving heart our only "inner compass." And she concludes, "If we want to be at home on this earth, even at the price of being at home in this century, we must try to take part in the interminable dialogue with its essence" (1953, 392). If we want to be at home on this earth, perhaps we cannot be at home in this century if it means alienation from both earth and world. Indeed, thinking with the earth requires thinking beyond this century and imagining the future of the planet and of our world(s). Furthermore, perhaps the possibility of being at home is bought at the price of continual wandering through the deserts and oases of our imaginations and of our planet. Not an aimless wandering, but a journey whose goal is to create a world in which we love the earth enough to take responsibility for it. This would be a world of sustainable ethics in which we limit ourselves in order to protect the literal and metaphorical ground of any possible world, which is the shared planet Earth.

4

THE EARTH'S REFUSAL

Heidegger

Although in Heidegger's corpus *earth* becomes a central theme in a way that it is not for either Kant or Arendt, the meaning of *earth* in his writings is far from obvious. Just as it did for Kant and Arendt, the concept of earth operates as a limit concept for Heidegger. Whereas for Kant, and for Arendt, a limit is the end point of our understanding or experience, however, for Heidegger, it is what makes experience and understanding possible. The limit is beginning rather than end (e.g., Heidegger 2000c [1935], 208). Whereas Kant saw the Earth as one planet among others, which one day we may inhabit, Arendt and Heidegger insist that the earth is our singular home; and the dynamic of our earthly environs reminds us that it is not merely a planet. In fact, Heidegger contrasts the earth with the planetary, which comes to represent the dangers of totalization inherent in the technological worldview. Whereas the earth is a limit to our will to control and dominate, the planetary and global generate illusions of limitless technoscientific mastery. What Heidegger calls the "planetary imperialism" of the technological worldview risks destroying both world and earth (Heidegger 1977a [1938], 152). Only by acknowledging the earth as a multifaceted limit to technoscientific planetary imperialism can we hope to avoid this destruction. For Heidegger, safeguarding the earth requires acknowledging our own limitations as finite beings who are not only deeply dependent upon the earth but also profoundly relational creatures who belong to the earth.

Thinking through Heidegger's changing meditations on earth, especially in relation to world and planet, will help guide us toward an ethics of earth based on limitation rather than on planetary imperialism.

For Kant, the physical limitations of the earth limit our political relations. For Arendt, politics is born out of the plurality of the earth and our unchosen cohabitation. Heidegger's evolving notion of earth gives us an even deeper notion of our connection to the earth such that the relationality through which we exist, and through which all beings are constituted, is dependent upon the earth. Moreover, in these chapters, it is through an engagement with Heidegger's thought that the notion of earth takes an ethical turn. Although Heidegger is not known for developing ethics, and the dark cloud of Nazism hangs over his politics, the deeper and more primordial connectivity that he associates with earth opens onto a responsibility for care and nurturing, which he sometimes describes in terms of shepherding, stewardship, and protecting. These terms, however, may be too akin to managing earth's resources to be appropriate to the poetic dwelling that he eventually recommends. While Heidegger does not discuss our responsibility to the earth in ethical terms, his repeated invocations of being as a response to a call from elsewhere, along with his insistence on restraint, and the uneasy relationship between the uncanny (*Unheimlich*) and home, and finally his various remarks on *Dasein*'s responsibility to the earth, have a distinctively ethical tone.

This chapter traces Heidegger's changing conception of earth in relation to world, home, and technology, in order to make explicit the ethical dimension of our responsibility to the earth. The introduction of earth leads Heidegger from his early notion of Dasein as world-forming to his later notion of Dasein as dwelling. Rather than making the world, Dasein becomes a dweller tasked with caring for it. Dwelling is a mode of relating to the earth and to the world as protector and caretaker rather than creator or maker. The introduction of earth also leads Heidegger from the notion of Dasein as fundamentally homeless to Dasein as homecoming. And his shifting view of home and our relation to it opens onto a sense of belonging that goes beyond human experience and extends to other forms of life—perhaps even to inorganic beings. Once the earth enters what Heidegger calls *the fourfold*, rootedness on the earth becomes an alternative to the rootlessness of the modern technological worldview.

Earth comes to play a central role in Heidegger's thought insofar as it conceals itself from us, resists, and refuses our attempts to conquer it. In this way earth resists technological framing in the sense of mastering or managing. Eventually, when it appears in the fourfold, earth, along with sky, mortals, and divinities, both reveals and conceals beings and the meaning of Being, again providing an alternative to dominant technological modes of ordering beings and our relationships to them. The introduction of earth also signals Heidegger's shift further away from metaphysics and toward a nonsubjective and nonobjective phenomenology of experience as primarily relational. In addition, earth contributes to world in an intimate yet polemical relationship, a deeper sense of what it means for Dasein to be thrown into the world and projected toward its future. Earth is associated with the past as given, our *native ground* and rootedness in history; but the past and history given as open to interpretation and reinterpretation such that they are never fixed or static but always dynamic and relational. The earth is not only what is given, but also what is pregiven: everything we can do and think, and everything we are, is the result of our embodied relationship with earth. We are essentially earthlings, and as such we are formed out of, and in relationship to, earth. World, on the other hand, is associated with the future as an engagement with this past and history through which, in decisive moments, we can transform our experience or be transformed by it.

The introduction of earth into Heidegger's writings also changes his thoughts on home. Whereas in his early work it is clear that philosophy is a sort of homelessness and authentic Dasein is not at home among beings even as Dasein is in the world, in his later work Dasein's rootedness in and on the earth brings a sense of home. Rootedness on the earth is home not in the sense of the nostalgic longing for some fantasy of home as presence, oneness, or unity, but rather rootedness on the earth is home in the sense of belonging to one's surroundings, both in terms of time and space, which is renewed through careful attention and watchful vigilance and, perhaps most important, thoughtful gratitude.[1] This sense of home as belonging rather than possessing signals a shift away from technological modes of ordering that make home a belonging as possession and the base sense of home as ownership. Dispossessed in relation to earth, sky, mortality, and divinities, Dasein itself becomes a homecoming, a transition, something that happens in between where worlds collide.

The ambiguity of Heidegger's notion of *Heim*, which can mean both inhabit and home, complicates Kant and Arendt's notions of inhabiting the earth as sharing the same planet, particularly insofar as cohabitation is thought merely in human terms and not in terms of nonhuman animal habitats.² Sharing the limited surface of the earth, which for Kant becomes the ground of public right, pales in comparison to Heidegger's more robust conception of inhabiting the earth. Inhabiting the earth is more than merely sharing its limited surface. And while Arendt adds the critical components of the unchosen nature of this sharing, along with cohabitation as something that is given rather than made by us, and plurality as the law of the earth, her conception of habitation is still shallow compared to Heidegger's images of roots and rootedness in earth as *Heim*. For Heidegger, habitation and home are given, but they are always given only insofar as they are also relational and constantly transformed by our attention to them. Taking it one step further, we could say that the "given facts" of the limited surface of the earth, and that we do not choose with whom to inhabit the earth, must be interpreted and reinterpreted, avowed and taken up, to assume ethical and political significance. In order to avoid the totalizing and homogenizing universalism of global thinking, which both Arendt and Heidegger warn against, we must rethink the earth and earthly existence outside globalism, seen as the leveling of markets and mass telecommunications. The question, then, posed by Heidegger's thought, is how to formulate an ethics of the earth that is not simply globalism, as in the globalization of technology. Or, in more Heideggerian terms, we might ask how the earth, and thinking with and against earth, might provide an alternative to technological enframing with its inherent globalism. How can we think the earth as opposed to the globe or the planet? How can we think the earthly as an antidote to the global and planetary? Moving beyond Arendt and Heidegger, can we reimagine a globalism based on the earth that opens up rather than closes off ethical and political responsibility?

BEING-IN-THE-WORLD

Given that much has been written already about Heidegger's early notion of being-in-the-world, and that our concern is the relation between world and earth, we will only briefly consider Heidegger's conception of world prior to the introduction of the concept of earth.³ For our purposes, suffice it to say that in *Being and Time* Heidegger insists that both the everyday notion of world as "the totality of beings which can be objectively present within the world," or as a "region" as in "the world of mathematics," already presupposes a more original sense of *world* as "that 'in which' a factical Da-sein 'lives'" (Heidegger 1996a [1927], 60–61). The world, or what Heidegger calls the "worldliness" of the world, is preontological insofar as everything that appears to us does so because we already are in the world. He describes the world as "a kind of being of Da-sein" and not a collection of objects, or the totality of objects, which exist. Later, in "The Age of the World Picture," he says, "the concept of world as it is developed in *Being and Time* is to be understood only from within the horizon of the question concerning 'openness for Being' [Da-sein], a question, that for its part, remains closely conjoined with the fundamental question concerning the meaning of Being (not with the meaning of that which is)" (Heidegger 1977a [1938], 141). World, then, is not a metaphysical category designating *what is*. Rather, world is a *way of being* that is particular to Da-sein. While other beings exist, only Da-sein is being-in-the-world.

Indeed, Heidegger uses *Da-sein*, literally, "there-being," rather than human being, at least in part to distinguish a mode of being in the world from a species of being.[4] In this sense Da-sein signifies that human beings have a sense of themselves as *there* in the world, a sense of their *there-being*. But, Da-sein *finds* itself there, and does not put itself there, because Da-sein is thrown into a world not of its own making. "To being-in-the-world belongs the fact," claims Heidegger, "that it is always already thrown *into a world*" (Heidegger 1996a [1927], 179). This means that there is another sense in which Da-sein is being there, namely as always already ahead of itself or, as Heidegger says, "*being-ahead-of-itself-in-already-being-in-a-world*" (Heidegger 1996a [1927], 179). In this sense Da-sein is always there rather than here. Da-sein is never here in the sense of present to itself. Rather, it is always running ahead of itself. This running ahead of itself becomes spatial as well as temporal in Heidegger's example of walking to the door in "Building, Dwelling, Thinking." He says that before we even begin walking toward the door, we already imagine ourselves at the door; before we take a single step, we are necessarily already at the door in our thinking about going there; we are already ahead of ourselves. In this regard, we are already *there* and not simply *here* (Heidegger 1971a [1951], 359). Da-sein is radically split between here and there, a being never fully at home with itself or in the world. The feeling of being at home is characterized as an illusion of the inauthentic and unthinking masses or what Heidegger calls "the They." Authentic Da-sein, on the other hand, realizes that its primordial condition is "not-being-at-home" (cf. Heidegger 1996a [1927], 177).

Already in *Being and Time*, however, we see tensions between Da-sein's thrownness into a world not of its making—its "falling prey" to "the They" and the "world"—and its structural relationship to the world as one of projection or always already being ahead of itself. The world into which Da-sein is thrown is not only the world of the past and traditions into which it is born, but also the world of its future and of the futural structure of its very being. How does Da-sein escape from the world of The They and attend to, or care for, the world as such? It is this in-betweenness that comes to dominate some of Heidegger's subsequent work. In between a past that it inherits and an open future, Dasein is never completely present to itself. In terms of *Being and Time*, this movement also corresponds to Heidegger's distinctions between our inauthentic relation to the "world" (which Heidegger indicates using quotation marks) of The They and an authentic relation to the world as such. It is important to note that at this point there is still an intentional subject (Da-sein) however passive, encountering an object (the world) however active, through care or attunement. As we will see, however, later Heidegger introduces the concept of *Earth*, and *strife* between earth and world, to further his quest for a nonsubjective, nonmetaphysical account of Da-sein's relation to Being as he moves further away from Husserl's notion of the transcendental subject's intentional relation to objects.

A year later, in "On the Essence of Ground" (1929), Heidegger again considers the meaning of *world*. Here, he links Dasein's projection, or always-already-ahead-of-itself structure, to transcendence—but a strange type of transcendence—which is both before and after the manifestation of beings as such because Dasein is world-forming, but never creates the world ex nihilo (Heidegger 1998b [1929], 123). Again, Heidegger attributes a peculiarly passive subjectivity to Dasein insofar as it not only lets the world occur but also gives itself the world; world-forming encompasses both the passive and active senses of forming. Obviously, Heidegger is struggling with the tension between the passive letting the world happen and the active giving itself the world. At this point in his thought, however ambiguous, Dasein is clearly still world-forming. My argument is that the introduction of earth helps Heidegger resolve this ambiguity by linking the passive happening to earth and the active giving to world. Later still, Heidegger lets go of the language of world-forming in favor of dwelling, also seemingly the result of the introduction of earth into his philosophy.

The world-forming essence of Dasein becomes Heidegger's primary focus in his lecture course, *The Fundamental Concepts of Metaphysics* (1929–1930). There, Heidegger reminds us that he has already addressed the question central to this lecture course—"What is world?"—in both *On the Essence of Ground* and *Being and Time*. He describes three "paths" toward clarification of the question: First, there is "the history of the word 'world' and the historical development of the concept it contains," which he pursued in *On the Essence of Ground*. He summarizes his findings, concluding that man is both part of the world and stands over against the world, both a servant of the world and a master of the world, which "does indicate man's ambivalent position in relation to the world—as well as the ambivalent character of the concept of world itself" (Heidegger 1995 [1929–1930], 177). My hypothesis is that Heidegger eventually introduces the notion of earth in order to clarify this ambivalence by dividing into two the two parts of man's relation to world, the passive part that he serves, namely the world that he inherits, which becomes associated with earth, and the active part that he builds, namely the world that he creates.

In contrast to this first historical path, Heidegger says that in *Being and Time* he took a second phenomenological path "by interpreting the way in which we at first and for the most part move about in our everyday world" (1995 [1929–1930], 177). And now he states that he has chosen to follow a third path, "the path of a *comparative examination*," which leads him to his famous three theses: 1. "The stone (material object) is *worldless*; 2 the animal is *poor in world*; 3. Man is *world-forming*" (1995 [1929–1930], 177). Man *has* world; this is his relation to it. Stones clearly do not have world; while animals are in between stones and humans, possibly having world but a poorer version of it. Since, along with many others, elsewhere I have discussed Heidegger's three theses in terms of animals, our focus is on the relationship between world and home as it pertains to humans in the hopes of further motivating Heidegger's introduction of earth five years later.[5]

PHILOSOPHY AS HOMESICKNESS

In *The Fundamental Concepts of Metaphysics*, Heidegger quotes Novalis, "Philosophy is really homesickness, an urge to be at home everywhere," from which he concludes that homesickness is "the fundamental attunement of philosophizing" (1995 [1929–1930], 5). Once we begin asking the question "why are we here?" we are displaced from home. Indeed, philosophizing makes it impossible to go home, to find home, if that means beings at home everywhere. In a sense, the rest of Heidegger's work is about orienting ourselves on the proper path to find our way home to Being; it is about thinking as thanking by way of the path toward home. Yet we are driven on this quest by homesickness, the uncanny (*Unheimlich*) feeling of not being at home, not being here.[6] Finding our way home cannot be mindless shuttling as in commuting home from work, neither is it an arrival once and for all. Rather, it is the dynamic relationality through which we belong to the world and to the earth.

What does it mean, then, for philosophizing to want to be at home everywhere? Heidegger answers, "Not merely here or there, nor even simply every place, in all places taken together one after the other. Rather, to be at home everywhere means to be at once at all times within the whole. We name this 'within the whole' and its character of wholeness the *world*" (1995 [1929–1930], 5). To be at home everywhere means to be at home in the world as such, not to be at home in a place, even if that place is one's home. "This is where we are driven in our homesickness: to being as a whole. Our very being is this restlessness. . . . We ourselves are this underway, this transition" (5–6). Human beings as Dasein are a transition, a coming and going home, and eventually a homecoming. Thus it turns out that in his comparative analysis it is not animals that are in-between; rather, humans are in-between. We might even say that if humans are never at home in the world because we are world-forming then we are even poorer in world than the animals. Presumably on Heidegger's account, animals at least are at home.

But what are humans in between? There are three terms in this comparative analysis, stones, animals, and man. Certainly humans are not in between stones and animals. Rather, it seems that humans are between animals and gods. Indeed, later in "The Letter on Humanism" (1946), Heidegger claims that we have more in common with gods than with animals, which are the strangest to us because they are akin to us and yet separated from us by "an abyss" (1993 [1946], 230). Heidegger suggests that the living creatures all around us are more uncanny than the gods (230). The human's relation to divinity may indicate another reason why Heidegger introduces the concept of earth and eventually of the fourfold (earth-sky, mortals-divinities). For, on the one hand, his comparative analysis seems incomplete without gods insofar as it cannot explain the in-betweenness of human beings and, on the other hand, the alien nature of other living creatures is not captured by the concept of world. Although Heidegger denies that his comparative analysis does not yield a hierarchy of beings because animals and humans are of such different orders that there is really no comparison possible after all, it seems that he comes to realize that he needs another term to distinguish other living beings from Dasein because relating them both to one and the same world suggests a kinship that overshadows the abyss between them.[7]

Heidegger suggests that it is our finitude that accounts for our in-between status and therefore our uncanny homelessness. It is through our finitude that we become individuated and unique and therefore alone. Each human being is unique because each is finite. And our finitude results in our solitude. But man's uniqueness, his finitude, is also his solitude. Heidegger claims, "Individuation—this does not mean that man clings to his frail little ego that puffs itself up against something or other which it takes to be the world. This individuation [*Vereinzelung*] is rather that solitariness in which each human being first of all enters into a nearness to what is essential in all things, a nearness to world. What is this solitude [*Einsamkeit*], where each human being will be as though unique [*Einziger*]?" (Heidegger 1995 [1929–1930], 6). *Einsamkeit* and *Einsamung* can mean either *loneliness* or *solitude*. And we may wonder if, like Adam alone with the animals before the creation of Eve, man is lonely because of the abyss between himself and other animals. Man is lonely, as John Berger maintains, insofar as he sees his species as alone, in both the sense of lonely and of unique; man is lonely because he is alone of his species (Berger 1991). This passage leads Derrida to ask whether or not the beasts are alone (Derrida 2011, 5). It seems that for Heidegger the beasts are not alone and neither are stones. They are not individuals in the way that human beings are individuals. And therefore they are neither without others of their kind nor unique, whereas each human being is *as if alone,* without others of its kind, and each one is unique. Perhaps this loneliness contributes to the homesickness and nostalgia that Heidegger associates with Dasein's uncanny way of being in the world. As we will see, in a sense, we are lonely and alone because other creatures and the earth refuse to speak to us.

For Heidegger, unlike animals or stones, we are "gripped" by philosophical questions about our existence. And this being gripped rips us out of any comfortable sense of being at home and throws us into the uncanny homelessness that is Dasein's way of being in the world and thereby results in homesickness. As we have seen, he concludes that philosophy itself is a form of homesickness or, more precisely, a demand to be at home everywhere at once. This, of course, is impossible and leads to homesickness since no one can be at home everywhere. Philosophy's demand, then, is not only impossible but also leads to a profound homesickness, which Heidegger identifies as the fundamental mood of philosophy (Heidegger 1995 [1929–1930], 7–8). And a mood is not something that we control. Rather, just as sleeping and waking are not the same as consciousness and unconsciousness, attunement is not the same as consciousness; attunement is more like awakening (Heidegger 1995 [1929–1930], 59–60). Just as we do not control sleeping or waking, we do not control attunement or being gripped; it is something that happens to us. Philosophy's grip makes everything familiar seem unfamiliar and everything homely seem unhomely or *unheimlich*, uncanny. Philosophy's attention to, or questioning of, *world as a whole* makes the world as the totality of beings disappear; in this regard, Heidegger links philosophy to *boredom*. Through what he calls *profound boredom*, the world disappears and we are left with the empty structure of Being, essentially time or the stretching out of temporality, which he describes as an emptying of the world: "Being left empty [*Leergelassenheit*] as Dasein's being delivered over to beings' *telling refusal* of themselves as a whole" (137, my emphasis). He describes this "telling refusal" [*Versagen*] as "in itself a telling [*Sagen*], i.e., a making manifest" and he asks, "What do beings in this telling refusal of themselves as a whole tell us in such a refusal? What do they tell us in refusing to tell?" (140). He answers that the refusal of access to the whole of being tells Dasein, through profound boredom (presumably when no single object engages attention), that there are "unexploited" possibilities (141). The refused possibilities of engagement announce other possibilities for Dasein. Furthermore, profound boredom interrupts any intentional consciousness of an object by a subject. In profound boredom there is neither subject nor object, but rather the emptiness of the whole as such. This is to say, profound boredom is an experience of ex-stasis that removes us from both the subjective and objective worlds and puts us face-to-face, so to speak, with the refusal of the world, the very indifference of the world as a whole. In other words, we are face-to-face with the refusal of the world as such to show its face, which means that we never come face-to-face with the world as a whole.⁸ While there is much more to say about Heidegger's notion of profound boredom, for my purposes here we will focus on the role of beings' "refusal" and its relationship to the earth's refusal, which appears later.⁹

This "telling refusal" is related to the revealing/concealing structure of truth as *aletheia* that Heidegger develops throughout his work, which is central to his introduction of the term *earth* as a counterbalance to *world* in "The Origin of the Work of Art" (1935–1936). Through the retreat from the world that happens to us and grips us in philosophical questioning, there is an emptying out of the content of the world and a focus on our relationship to it. Once we attend to the world and our relationship to it, we realize that the world always recedes from our grasp; indeed we are gripped by it. Moreover, the world reveals itself, or is revealed, as always necessarily concealing itself, akin to what Heidegger characterizes here as the "telling refusal" of being.

It is revealing that in *The Fundamental Concepts*, Heidegger uses the same term, *Versagen*, to describe the animals' refusal of "being with": "what is it about the animal which allows and invites human transposedness into it, even while *refusing* [Versagen] man the possibility of going along with the animal? From the side of the animal, what is it that *grants the possibility of transposedness and necessarily* refuses *any going along with*?" (Heidegger 1995 [1929–1930], 210; my emphasis). Like being's refusal in profound boredom, the animal's refusal is also a telling. *Versagen* is also *Sagen*. And what the animal's refusal tells Heidegger is that the animal's world is limited and poor insofar as it does not have the ability to transpose itself, while humans do. For Heidegger, animals are not capable of the *Mitsein* or *being with* that is characteristic of Dasein. He argues that human beings can transpose themselves into others; most simply, they can put themselves in the place of others (205). Animals, on the other hand, are incapable of such transposition. This leads him to ask whether or not we can transpose ourselves into animals. His answer seems to be "yes and no." We project our own world onto animals, especially our domestic pets. Yet, according to Heidegger, we live in separate worlds that are radically inaccessible to each other. Indeed, for him, it is questionable whether animals have a world at all. They have and yet do not have world.¹⁰ What is noteworthy is that animals refuse us and not the other way round, and this is what makes them so uncanny for us: akin but separated by an abyss.

Animals refuse to tell us, to say, whether or not they have world or what their world or experience is like. And because they refuse to tell us, our ability to transpose ourselves into their "world" is limited. Furthermore, in some sense, it is this limitation—that is to say, *our own* limitation—that leads Heidegger to conclude that animals are poor in world, that they have limited access to world. Because animals do not speak and tell (Sagen) us about themselves, we cannot know them. Because human beings are speaking beings, we exist in language and therefore can transpose ourselves into others. So, too, because beings as a whole are speechless, they refuse to tell us about themselves. Thus their being remains concealed from us, even as it is revealed as concealing itself or as refusing us. The *unheimlich* or uncanny experience of man is in part the result of living amongst so many beings that refuse to speak to us. In this sense we are alone and therefore also lonely. We are not at home with the "silence" of other beings. And the danger is that, rather than listen to their telling refusal, we try to make them speak, force them to tell us their secrets, the secrets of being. This is why Heidegger insists on letting beings be, *Seinlassen*, rather than mastering them and why, later, he talks of listening to the earth.[11]

EARTH IN STRIFE WITH *WORLD*

When Heidegger develops the notion of *earth* in "The Origin of the Work of Art," he also associates it with concealment as refusal.[12] There, he discusses two types of concealment, refusal and dissembling, and ultimately we cannot tell them apart (Heidegger 1971c [1935–1936], 179). We don't know whether the earth (or animals) is refusing to tell us or whether it is dissembling. In the following few years, in his notes in *Contributions* (1936–1938), Heidegger continues to link earth to refusal, particularly refusal as the essence of event or the happening of truth. In an important sense the event of truth is the earth's refusal (Versagen) to speak to us. This means that our way of being on the earth, the essence of our way of being, is as limited beings for whom the truth is concealed. What is revealed in our relation to earth is precisely this concealment at the heart of the essence of truth, which is to say the earth's refusal. *If there is an ethical imperative in Heidegger's thinking, it is the imperative to listen to this refusal.* We must hear and acknowledge the refusal as the abyss between us and animals or earth.

Not in order to deny our relationships to them. On the contrary, to insist that we exist as fundamentally relational creatures that belong to an earth we cannot conquer or control. Therein lies the ethical import of Heidegger's thought. Namely that we must hear and heed the earth's refusal and allow it to appear as such. In one sense this is the seemingly simple lesson that life is unpredictable and that we cannot control or master it. Yet figuring out how to heed this simple lesson is difficult. What, then, is the appropriate response to the earth's refusal?

Still trying to wrest free of the traditional metaphysical distinction between subjects and objects, Heidegger invokes what he calls the *strife between* earth and world as beyond either subjects or objects. His interpretation of the work of art takes us out of either subjective or objective theories of aesthetics and toward the truth of the work of art as event or happening [*Ereignis*]. Although in "The Origin of the Work of Art" Heidegger associates clearing or opening with world and refusing or concealing with earth, he is explicit that this division is too simplistic. He is not describing two distinct realms or spheres, but rather a dynamic relationship through which each comes into its own as a process and not a substance or object. He says, "the world is not simply the open region that corresponds to clearing, and the earth is not simply the closed region that corresponds to concealment. . . . Earth juts through the world and world grounds itself on the earth only so far as truth happens as the primal strife between clearing and concealing" (Heidegger 1971c [1935–1936], 180). Concealing and unconcealing are always together in the happening of truth as the strife between world and earth. World and earth do not exist apart from each other, but only in their conflict, the essential conflict between opening and closing, revealing and concealing, *Sagen* and *Versagen*. Yet Heidegger also argues that earth reveals itself as concealing, and in this way the earth withdraws. Although earth and world appear as the two elements engaged in struggle, Heidegger's emphasis on the dynamics of strife and the intimacy between the two elements makes clear that they are necessarily dependent on each other. Indeed, their very being is possible only through their relationship; they do not exist separately.[13]

Just as the world reveals itself as revealing, earth reveals itself as concealing. Yet, neither earth nor world is an object that stands before us; rather, both come to be through dynamic processes of relationship. This is why Heidegger says, "the world worlds" (Heidegger 1971c [1935–1936], 170).

Here Heidegger insists that the world is neither "the mere collection of the countable or uncountable familiar or unfamiliar things that are at hand" nor "a merely imagined framework added by our representation to the sum of such given things" (170). The world is neither the realm of objects and objectivity nor the realm of subjects and subjectivity. The world is neither outside us nor inside us; it is neither Kant's noumenal world nor his phenomenal world that is the result of categories of our own understanding. Rather, the world is where we find ourselves, a being amongst beings and yet a being uniquely related to its world. Here Heidegger repeats his claim that the stone is worldless, while human beings, even the peasant woman, have a world (170). It is important to point out, however, that at this point Heidegger still suggests that human beings are not at home in the world but merely believe themselves to be at home when they are not attending to the truth of Being (cf. 170, 179). We will return to "home" later with the question of whether or not earth is our home and whether or not we are at home on earth.

By *earth*, Heidegger clearly does not mean the planet: "What this word [*earth*] says is not to be associated with the idea of a mass of matter deposited somewhere, or with the merely astronomical idea of a planet.... Earth occurs essentially as the sheltering agent" (1971c [1935–1936], 168). But what does earth shelter? It seems that earth shelters unconcealment and continually reminds man that he is not the master of Being. Heidegger claims, "Earth thus shatters every attempt to penetrate it. It causes every merely calculating importunity upon it to turn into a destruction" (172).[14] For example, when we measure the wavelengths of light, we miss the shining appearance of colors; when we calculate the density of rocks, we miss the heaviness of their weight and the texture of their surfaces.[15] Thus, when we attempt to master the earth and penetrate it through technoscience, we risk destroying it. We destroy its sensuous appearing as shining through, which to say our experience of it. We also risk literally destroying the earth, not only physically, through the threat of nuclear destruction, but also, and moreover ontologically, through the illusion of control inherent in the technological worldview. This is to say, that the technological worldview denies that the earth is essentially undisclosable and not only attempts to unlock its secrets but also is confident that in principle it can know everything there is to know about the earth. Paradoxically, the earth appears as itself when it appears as concealed and withdrawing from view.[16] And we are obligated to listen to its refusal and acknowledge its essential withdrawal. The earth's essence is refusal, concealment, and withdrawal, and that fundamental limit is what keeps our dangerous illusions of mastery and planetary imperialism in check.

The earth is that which prevents man from bringing everything into his world and into the open. The earth is "self-secluding" (173). The way that earth makes its way into world is as self-secluding or as a telling refusal that shows us that human beings can never understand all there is to Being. The earth always withdraws from view within the world. In this way earth is a sheltering as both protecting and hiding the truth of Being from man's violent attempts to dominate the earth through technoscience. In addition, Heidegger describes the earth's sheltering as the ground of "historical man's" dwelling in the world (171–172). As we will see, when he turns his attention to the meaning of *dwelling* years later in "Building, Dwelling, Thinking," earth enters into *the fourfold,* and the relationship between earth and world takes yet another turn in his thought.[17]

If the earth is the ground of the dwelling of historical humans, the world is the self-opening of "essential decisions in the destiny of a historical people" (174). This tension between the history of humans and their destiny through essential decisions is the strife between earth and world. In this regard, "the world, in resting upon the earth, strives to surmount it" (174). We can never escape, or even completely understand, our own history as individuals or as a historical people, let alone the history of humans—as if humankind in general has a history. It is what is given and therefore essentially unknowable and, because unknowable, necessarily a matter of interpretation. Thus it is never given once and for all, but always open to reinterpretation. In this sense the givenness of our history is unchosen and yet must be chosen. Although we cannot change our history, we must decide how to take it up. And it is in that space between the unchosen and the chosen that ethics happens. This is the space between earth and world.

As earth, what is given, or history, is always concealing and refusing, while as world it is always revealing and telling (200). As world, our transformations of earth bring us into our destiny as individuals and as a people. We attempt to change our history and move forward through our essential decisions both as individuals and as historical peoples (and perhaps even as human beings more generally, although, for Heidegger, claims to this kind of universality, the universality of Kantian perpetual peace, are not only suspect but also dangerous). Concealing and revealing, past and future, history and destiny, ground and decision, the "ordinary" and the "awesome" come together through the strife between earth and world (cf. 201). The world can interrupt the earth and history begins again; then a new world irrupts (201). So, too, the earth can jut up through the world and appear as the sheltering of concealment that prevents us from illusions of mastery and dominance over the totality of beings and Being as such. The strife between earth and world is the "intimacy with which opponents belong to each other" (180). Earth and world belong to each other. And truth happens between them, in their relationship, the relationship between the unchosen and the chosen, the given and interpretations of it.

In terms of the work of art, earth resists and refuses ever being used up in any one representation or interpretation. If the figure of earth is one of concealedness, which shows us that being is always open to interpretation and reinterpretation even while resisting or refusing any one final totalizing fixed universal or eternal truth, then we might say that Heidegger's introduction of the notion of earth performs or enacts that operation as well as announcing it.[18] For, perhaps more than other enigmatic concepts in the Heideggerian corpus, earth has been open to surprisingly divergent interpretations, not only when it later enters the seemingly mysterious mythical fourfold but also here in its strife with world.[19]

In "The Origin of the Work of Art," the earth as "native ground" is self-secluding and always withdrawing from view such that what is given is essentially unknowable and therefore must be taken up through poetic interpretation into the world.[20] Through the strife between history as given (earth) and history as made (world), we are simultaneously rooted and uprooted in our interpretative existence.[21] This history, the history of a people, is never fixed; rather, it is always a matter of interpretation and reinterpretation.[22] Yet, the Earth, and our earthbound existence, always precedes every interpretation as its forestructure and as foregiven. Moreover, this foregivenness is what enables every interpretation, including scientific studies of the Earth itself (cf. Fried 2000). As such, the Earth as foreground always necessarily recedes from our grasp since it is the ground of every attempt to grasp it. It is what makes all attempts to grasp possible. We might call this the pregiveness of earth. Not only is earth associated with the past and history as unchosen, but also the earth is the stuff out of which we are made, physically, emotionally, mentally, and completely. We are of the earth in a way that determines our being. Resonant with Husserl's insistence that the earth is a basis body or primordial arc, and Arendt's dogged reminders that everything we do is of the earth, Heidegger's earth is the ground of our very being.

We can never step outside of that pregiveness in order to take it as our object. Rather, everything we are and everything we do or think is of the earth. And yet there is a real and present danger that we come to believe that we can create a world either as a prosthetic substitute for earth or in order to control it.[23] Earth and world must remain in productive tension, otherwise we risk the danger of, on the one hand, viewing our history as our destiny such that it becomes unconditional and an end in itself, as it did with the National Socialists in Germany and other nationalist movements, or, on the other hand, fall under the illusion that we can escape our history entirely and make ourselves anew ex nihilo, as with some technologies that are presented as mastering nature and human destiny by some bioethicists such as John Harris in their claims for genetic engineering.[24] The former is the triumph of earth over world, the past over the future, while the latter is the triumph of world over earth, the complete disavowal of history. And both shirk the responsibility of interpretation as the call to thinking as questioning and thus of strife. Indeed, both act as if earth and world are static and transparent to us in such a way that they do not require interpretation. We have an obligation to continue to reinterpret our past for the sake of the future and of our world.[25]

As is becoming evident, in Heidegger's corpus neither earth nor world is static or transparent to our understanding; rather, both require ongoing interpretation and reinterpretation.[26] Moreover, both are necessary for interpretation. Earth and world, and more precisely the strife between them, is what both necessitates and engenders the possibility of interpretation. Perhaps not surprisingly, Hans-Georg Gadamer, for whom hermeneutics is primary, insists that earth is not material or stuff, but rather the depth of meaning itself out of which everything emerges.[27] Gadamer describes the dynamic tension between earth and world as the tension between lightness and darkness, both of which are necessary for life and meaning: "It is not only the emergence into the light but just as much the sheltering itself in the dark. It is not only the unfolding of the blossom in the sun, but just as much its rooting of itself in the depth of the earth" (1994 [1960] 106). The metaphor of the flower blossoming in the sun but taking nutrients in the darkness of the earth beautifully shows the interrelation between light and dark, sun and earth, or, in Heidegger's more abstract terms, world and earth. The world is the drive to bring everything into the light, while the earth shelters the roots in the dark. And both are necessary for life.[28] In terms of history, we could say that the tension between the artwork's rootedness in tradition and its originality or newness gives it its power. This is the strife between earth as history and world as innovation or what we make of that history, which does not mean that we control either. If earth is the force of the past, world is the force of the future. And we could say that the present is always the shifting effect of the tension between the two.[29] The power of the work of art derives from the meeting of dark and light, earth and world, past and future, given and made.

Heidegger emphasizes the in-between, the relationship, the strife between earth and world. This between or relationship is the excess that takes us beyond the dualisms, subject or object, earth or world, past or future, and into opens up the possibility of creativity and innovation. The work of art emerges from this between or relationship and not from one term or element or the other. Indeed, Heidegger insists that one does not exist without the other. The tension between them is the productive tension that allows us to interpret and reinterpret our world endlessly, just as we might a great work of art. Heidegger's introduction of the term *earth* as counterbalance to *world* signals his attempt to wrest free from both subjectivism and objectivism and propose relationality or in-betweenness as the locale of truth.[30]

The work of art is the quintessential example of this tension between earth and world, past and future, old and new, given and made. The great work of art must both follow traditions and yet create something new. Through the work of art, the earth or history and tradition jut through and yet are interrupted by the world as projection into the future. The earth as what is given, which is to say, our history as individuals and as peoples, "juts through the world," without our complete understanding or permission. But, even as the "earth juts through the world," "at each time a new and essential world" irrupts (Heidegger 1971c [1935–1936], 174, 201). The world, or worlds, which we create, not through our individual will or mastery of earth, but rather through listening to the earth and interpreting our history, irrupts onto the earth, each time anew. Works of art become great works insofar as they open onto a world and reveal the earth as self-secluding or withdrawing, inviting interpretation and yet refusing any definitive truth that silences the call for questioning and thinking. For, as Heidegger says at the end of "The Question Concerning Technology," "questioning is the piety of thought" (Heidegger 1977b [1954], 341).

RESTRAINT, REFUSAL, AND LIMITATION

It becomes clear from Heidegger's remarks in *Contributions* (Heidegger 2012b) that earth, and the strife between earth and world, are associated with restraint, and listening to the earth requires restraint. There he suggests that restraint is the style of preparedness for the event of truth (§13, 29). "This restraint also pervasively disposes all playing out of the strife between world and earth" (§31, 55). Restraint is a way of preparing for the strife between earth and world, and it disposes the essential strife between earth and world. Restraint and stillness give rise to truth insofar as they let it happen. Heidegger says, "Great stillness must first come over the world for the earth. This stillness arises only out of restraint" (§13, 29). This passage suggests that the world must become still and restrained for the sake of the earth, for the sake of the strife between world and earth, and that, by restraining and stilling, the world as the appropriation of the truth of being can happen. What is revealed through restraint and stillness is that there is always something essential concealed and, furthermore, that concealing is itself the essence of truth as happening. If the drive of world is to reveal, then balance must be restored in the strife between world and earth. World cannot dominate earth or we risk falling under the illusion that we can predict and control both earth and world by bringing everything into the light. Switching to Heidegger's auditory metaphor, only by silencing the chatter of world and worlds can we listen to the self-secluding refusal of earth. Furthermore, only by stilling the desire to control and to bring everything into the light, which is associated with the world, can we open ourselves to the earth as the preservation of the refusal of this illumination. But what must be restrained in order for the world to be stilled? The will to power of human beings must be restrained. The human drive to penetrate and pry open every corner of the earth, to bring everything into the light, must be restrained in order to listen long enough to dispel the fantasy that in principle, if not in practice, some day we will know the secrets of the earth and of the entire universe. Only through the refusal of earth, and the restraint of the human, can we stay rooted on the earth. Once exposed to the light, the roots die, and so does the whole being. With the introduction of the concept of earth, we see Heidegger moving away from the notion of Dasein as actively world-building and toward an emphasis on restraining Dasein's building power.

Thus we have to restrain our desire to know. Moreover, we must guard the earth's refusal as a telling refusal. That is to say, the earth's refusal is not nothing; in its silence it tells us something, namely that it will never give up its secrets. In *Contributions* Heidegger describes Dasein's "proper role" as "the grounding stewardship [*Wächterschaft*] of the refusal" (§271, 383). It is the responsibility of Dasein to preserve and protect the earth's refusal to be penetrated by our own attempts at bringing it into the light of the world. Dasein's responsibility is to restrain itself. But this obligation cannot be carried out by sheer will power.[31] It is not a matter of overpowering ourselves in the same way that we try to overpower nature. And yet, what does it mean to "steward"? While the English word *stewardship* still contains an element of control and management, Heidegger uses the word *Wächter*, which has the sense of guard or watcher and avoids the connotations of management inherent in *steward*, especially when we think of the steward of an estate. *Wächter*, however, has strong associations with the military and force that are not present in the English *steward*. As such, it may not yet "capture" the cessation of agency and mastery necessary for guarding the earth.

Unlike the weaponized military guard, guarding or watching the earth, while a defensive position that connotes vigilance, is a matter of restraint and listening, a passive power that moves beyond any moral directives, universal principles, or dialectical resolutions that come from on high and are carried out through man's will. Perhaps this is why Heidegger talks of "the last god" and moving beyond all theism. He says that the last god is not just an end but also a new beginning. Possibly, then, the last god signals the end of fundamentalism and the beginning of true questioning. We must wean ourselves from fundamentalist belief in any god, in the sense of theism, universal truths, or dialectical resolutions and accept that we do not and cannot know or understand our own existence. As Socrates claimed, all that we know is that we do not know. Our task, as Heidegger describes it, is to steward or guard the earth's refusal to our own probing reason.

Provocatively, Heidegger asks, "Whether a human being is masterful enough both to withstand the resonating of the event as refusal and to carry out the transition to the grounding of the freedom of beings as such, i.e., the transition to the renewal of the world out of the saving of the earth—who could decide and know that?" (§256, 326). This passage suggests that saving the earth requires a renewal of the world and a renewal of the strife between earth and world; this renewal of world is beyond the fundamentalisms of either religion or science—the last gods. This renewal of world is such that it engenders the possibility of saving the earth. Conversely, the passage suggests that the renewal of world must come from saving earth. In other words, saving the earth may be possible only when we recreate the world with earth at its center, an anti-Copernican revolution. Perhaps this circular logic is but a symptom of the necessary strife between earth and world.

Heidegger asks whether human beings are masterful enough to carry out the transition to this renewal of world, which necessitates the self-restraint necessary to withstand the earth's refusal. Are human beings masterful enough to control themselves, namely to control their own will to control, long enough to heed the earth's refusal? And how can we control our own will to control without using the same logics and delusionary belief in our own inflated abilities that leads to the dominance of world over earth in the first place? The question with which Heidegger ends the passage suggests that we cannot *decide* or *know* how to save the earth in order to renew the world, or how to renew the world in order to save the earth, or which comes first, or how to enact the transition in ourselves necessary for that task. And, yet, if we can withstand the refusal of the earth and hear it as a telling refusal, then there is hope for renewal and a transition that might save the earth. To hear the earth's refusal as a telling refusal is to heed its refusal by giving up the illusion that we can master earth. The earth refuses to give up its secrets and we must not only hear its refusal but heed it in order to save the earth.

Heidegger suggests that only by freeing beings from our attempts to grasp them can we hope to save the earth from destruction. This freeing responsivity to the earth that sustains us is a safeguarding of each creature, each element, in such a way that all of them are freed to respond in their own ways. Freeing as letting be. Are we masterful enough to let beings go? This sense of freeing requires listening and attending to others and otherness rather than trying to control or master them. It requires letting the other be in such a way that it also fosters and nurtures. This notion of freedom is not based on autonomy but rather on the radical and fundamental relationality of each unique being as it participates in the gathering of earth and sky, mortal and divinity as beyond human reason or recognition, mastery or control. Heidegger insists that saving the earth will not come through human will power. On the contrary, to save the earth from our own destruction of it, we must carry out the restraint of human will power.

Only the refusal by earth to come into the light can counterbalance the persistence of the world's insistence on bringing everything to light. The earth's refusal is what ensures the "question-worthiness" of being because there is always more that we cannot know or understand. The earth's refusal (*Versagen*) also tells (*Sagen*) us that questioning is our proper mode of being insofar as it is essential to our way of being on earth. We are beings who question. Question worthiness "compels all creating into plight, erects a world for beings, and rescues what is reliable of earth" (§26, 50). The question worthiness of being turns creating into a quandary through which worlds come into being and what is reliable of earth is rescued. Questioning confounds attempts to secure mastery and reassure ourselves that we are the creative force between both earth and world. In English, in addition to the senses of quandary and confounding inherent in the word *plight*, there is also the sense of pledge or vow that connotes an ethical obligation to question, which is not present in the German word *Not*. However, there is a sense of urgency and necessity in the German that may not be as forceful in the English word *plight*. Questioning opens up the world and worlds such that alternative possibilities appear. By so doing, what is reliable of earth is saved. Since earth has been associated with refusal, we could say that what is reliable about earth is its refusal and that its refusal and its self-seclusion is what is rescued through questioning.[32] In other words, questioning recalls that there is always something essential that withdraws from our grasp, even as the question worthiness of being compels us to continue to ask and answer. Both world and earth are dependent upon constant and ongoing interpretation compelled by the question worthiness of being. Indeed, what Heidegger calls the strife between earth and world gives rise to interpretation, and, in a sense, interpretation itself could be seen as a form of strife.[33] Without the strife of interpretation, the strife between earth and world, not only do we risk losing the earth's protective self-seclusion that at worst leads to its destruction but also we risk losing the world whose meaning is anchored in the concealment of earth. In *Introduction to Metaphysics* Heidegger describes this strife in terms of the Greek *polemos*.[34] Interpreting Heraclitus, he says: "Where struggle ceases, beings indeed do not disappear, but *world turns away*. . . . Beings now become just something one comes across; they are findings . . . the present-at-hand, within which *no world is worlding*" (2000c [1935], §48, 65–66, my emphasis). Without struggle or strife—polemos—beings no longer can refuse us and become merely things at our disposal, for anyone to use. Even nature becomes natural resources and scenic views. Only the earth's refusal and our own restraint ensure that struggle continues. Strife not only gives rise to questioning but also ensures that it will continue. The question worthiness of being necessitates strife or polemos. Elsewhere, Heidegger insists that polemos does not mean war but rather strife, "not strife as dispute and squabbling . . . and certainly not as use of force" but as "being-exposed" (1985, 488). Without the strife or struggle with earth, beings in the world become nothing more than objects or findings, scenic views or pictures, and all at our disposal and under our control. Earth's self-seclusion and concealing is what prevents any one interpretation from becoming a dominant worldview that stops the world from "worlding"; such a worldview fixes both earth and world under man's dominating gaze.

Anticipating my discussion of Derrida in the next chapter, it is interesting to note that for Heidegger the world turns away when beings cease to be question worthy, when they no longer can refuse our attempts at mastery, and when everything is reduced to stuff at our disposal. The world turns away when technological mastery dominates our way of relating to earth. For Derrida, as we will see, the world turns away when the sovereignty of individuals and nations is interrupted by the ethical face-to-face that obligates one to carry the other. In a sense, for Heidegger, when struggle ends, the world turns away, while for Derrida, the struggle begins only when the world turns away. The question, of course, is what they mean by *world* and by its turning away. As we will see, at stake for both Derrida and Heidegger is the openness to interpretation and the cessation of the sovereign will, which both come to be associated with poetry. And for both, what is concealed is as important as what is revealed.

When human will dominates, the polemos between earth and world, concealing and revealing, is distorted into the fixity of the technoscientific gaze that overlooks the "in-between," the dynamic force of relationships as they are born out of strife, and, when this happens, we operate under the fantasy that revelation has triumphed over concealment, world over earth. Again Heidegger emphasizes the *between*, saying, "this 'between' as the ground of the strife of world and earth; history is nothing other than the eventuation of this 'between'" (2012b [1936–1938], §268, 377). History happens between earth and world, between past and future. And yet it is the earth, and the way that we take it up, which is in the most danger of being ignored. This is to say, within the technoscientific worldview, concealment is taken to be a temporary condition that can be eventually overcome with enough time and technology. Heidegger insists, to the contrary, that concealing is the way of the earth and our fundamental relationship to it.

In his postwar writings, Heidegger indicates that in our technological age one of the greatest challenges is attending to the way in which the technological worldview transforms our relationship with other beings, and being as such or Being, into planning, calculating, and managing. Taken to its extreme, this technological worldview leaves us with nihilistic machinations and no possibility for alternative relationships. This technological relation to world and earth is one-dimensional and reduces experiences to what can be quantified and used. Deeper meanings and the quality of life are threatened. Moreover, our questioning mode of being is overtaken by technological mastery, which leads to the illusion that our powers of comprehension are, in principle, endless. Without restraint, refusal, and limitation, we take ourselves to be the masters of Being and beings rather than stewards or caretakers, as Heidegger often claims. Even the notions of stewardship and caretaking, however, give the illusion that we are in control and may imply that we have a management position in relationship to something—the earth and its resources—that is in need of management. The technological worldview is not only alluring but also difficult to avoid.

Only by accepting earth's refusal to speak to us do we restrain ours attempts at domineering mastery; only if we listen to earth's telling refusal can we hear the need for interpretation and the impossibility of questioning coming to an end. Foreshadowing his later assessment of the dangers of technology, in *Contributions*, Heidegger warns that human beings "may for centuries still ravish and devastate the planet with their machinations, and the monstrousness of this drive may 'develop' to an inconceivable extent, assume the form of an apparent strictness, and become the measuring regulation of the devastated as such" so that, eventually, "all that matters is the calculation of the success and failure of the machinations" (Heidegger 2012b [1936–1938], § 255, 324). Technological machinations can take over both earth and world and leave us calculating every aspect of life in terms of the success or failure of our ability to calculate.[35]

Written around the same time as *Contributions*, "The Age of the World Picture" (1938) echoes some of the same sentiments about the dangers of the technological worldview (or any "worldview") taking over the earth. Here Heidegger sets out two possible paths. Either we preserve the horizon of concealment or we allow technology to usher in an unlimited sphere of objectification where everything is accessible and representable to everyone and binding for all (*1977a [1938]*, 147). Technology is a great equalizer insofar as it promises access for all. And yet, as we know, this promise covers over injustices and inequalities that result from the quest for more resources on the part of those who already have them. Certainly, globalization with its promise of bringing access to technology to everyone on the planet is a prime example of the way that inequity and injustice is hidden behind the fantasy of global technology. In truth, global technology and global markets have increased rather than decreased the gap between the richest and the poorest, those who have access and those who do not.[36]

Heidegger claims that the ideal of technological uniformity, which presumably includes the ideal of equal access to technology, "becomes the surest instrument of total, i.e., technological, rule over the earth" (*1977a [1938]*, 152). Everyone buys into this ideal as the promise for global unification. Writing before the language of globalization, Heidegger talks of "the planetary imperialism of technology organized man" (152). Inherent in the technological worldview is the planetary imperialism that attempts to dominate the entire globe using technology, and it succeeds in taking over the planet through the promise of equal access.[37] Yet the planetary in this technological sense is at odds with the earth as the sheltering agent that shelters concealment and withdrawal. For the goal of technological planetary imperialism is to penetrate every corner of the earth, both literally and figuratively, to both unlock and harness its natural energy and pry open its secrets.

Here Heidegger describes the modern age as one in which the world itself has become a picture that stands over and against the human subject. As the world becomes more of an object, man becomes more of a subject with control over it and vice versa (133). The leveling affect of technology, as we learn in "The Question Concerning Technology" threatens not only to turn the world and all of its beings into things standing in reserve for human use but also to turn human beings themselves into things standing in reserve. Only through reflection—thinking and questioning—can we preserve the horizon of unconcealment-concealment and take up our uncanny place among beings: "Reflection transports the man of the future into that 'between' in which he belongs to Being and yet remains a stranger amid that which is" (136). Reflection unsettles the technological worldview by returning man to this in between place where ambiguity reigns and questioning never ceases. The ambiguity of the human relationship to both earth and world resist the planetary imperialism of the technological worldview or world picture. For, although for us reflection comes through representation, its ground is of the earth and therefore nonrepresentational. Furthermore, while we attempt to understand our experience through representation, that experience always both precedes and exceeds any attempts at reducing it to representation. We occupy this uncanny place in between, the ambiguous space and time of being there and never here.

THE AMBIGUITY OF HOME

This brings us back to the question of home and whether man is at home in the world or on the earth or whether he is indeed homeless. Heidegger's position on the homely or homelessness, *Unheimlickkeit*, of Dasein seems to change dramatically throughout his work, particularly in connection to the role of earth in relation to world.[38] As we have seen, in his early work, he maintains that Dasein is uncanny, a stranger among beings, which makes philosophy a form of homesickness, a longing for an impossible home. Dasein is by nature ex-isting; Dasein is both in the world and yet can question its place in the world. By the time of *Contributions* and "The Age of the World Picture," Dasein's relation to home is even more complex insofar as Dasein is both at home and not at home in the world. With the addition of earth in "The Origin of the Work of Art," there is the added question of whether or not Dasein is at home on the earth, which leads to the question, if it is not a planetary body, what is this earthly home? It is noteworthy that Heidegger rarely says *the* world or *the* earth, but rather *world* and *earth*. That is to say, he rejects the definitive article when deploying these familiar terms in his new conceptual vocabulary. While world and earth may be decisive for man, they are far from definitive in any fixed sense. The fact that Heidegger employs various senses of "being at home" leads to a sense of ambiguity in man's relation to world and to earth as home. It is clear from his early work that in our everyday experience living amongst "the They" and engaging in our "idle chatter," we have a false sense of being at home. We feel comfortable in the world. As we begin to reflect on that world and our place in it, however, we lose that easy comfort and experience ourselves as uncanny, even homeless. This leads to philosophical homesickness and nostalgia for that easier life of being at home. At the same time, it makes us aware, in Heidegger's terms, that authentic existence requires philosophical reflection. In his later work, however, it seems that working through philosophical thinking toward poetic dwelling leads us again to being at home in Being insofar, as he famously says in "Letter on Humanism," "language is the house of Being. In its home man dwells" (1946, 217). Furthermore, the *Unheimlich* or uncanny and unhomely become associated not with man's being-in-the-world, but rather with the modern technological worldview that uproots man from earth and risks the turning away of world. No longer is homelessness an existential condition of Dasein, but rather the result of the technological mode of enframing that turns beings into objects and Dasein into a subject. And no longer is homesickness inherent to philosophy, but rather to the technological age that uproots us from earth, displaces us from world as worlding, and quells the strife between earth and world through penetrating fixity and illusions of control. Our ambiguous relation to home and homelessness suggests that our relation to home is itself filled with strife and that there is a false sense of being at home that comes from lack of reflection and unthinkingly taking on a fixed worldview without critical interpretation; then there is a true sense of being at home that comes from the dynamic process of ongoing interpretation as openness to beings as they appear through what earlier Heidegger called the strife between earth and world, past and future, the given and the made. This "false" sense of being at home is the lack of critical distance that comes with assuming that the technological worldview is our only relation to the world, whereas this "true" sense of home speaks to our ambiguous place in the world, even as we belong to the earth. We are on a journey, as Heidegger says, "becoming at home in being-not-at-home."[39]

The shifts in Heidegger's thoughts on home very roughly correspond to his changing notion of world, especially in relation to earth. As we have seen, in his earliest work, before he introduces the concept of earth, Dasein is fundamentally *unheimlich*; by having a world he is never at home in it. By his middle period, when he introduces the notion of earth, Dasein belongs to the earth, yet is uprooted from it through the objectification and subjectification that come with both the technological age and metaphysics. Dasein's in-between status makes Dasein both at home and not at home in the world and on the earth, which renders Dasein's existence fundamentally ambiguous. In *Contributions* Heidegger calls our ambiguous in-between home and not home, "refusing belongingness" (2012b [1936–1938], §270, 381). We belong to the earth even as it refuses us.

When Heidegger discusses being at home as dwelling, he associates home with guarding and caring. Home is not just a well-equipped space, rather it is a habitation that must be protected and preserved. Preservation is a process of questioning and interpretation of what has been given such that home is both unchosen and yet must be chosen. That is to say, although we did not choose earth as our home, we must choose earth as our home. Earth is "the 'home' which, though being completely spontaneously given, keeps asking to be chosen, adopted" (Haar 1993, 63). There is an ethical tone, an undercurrent of obligation, in our need to take up the earth as our home, to protect and preserve it through questioning and interpretation. The fact that we do not choose to live on earth makes it possible to choose to make it home. Because we have no choice, our choice becomes ethical. Because earth is given as our home, the ways in which we make it home are always matters of ethics. Making the earth home is not about carving out a space or populating it with buildings and houses, but rather it is about dwelling in every sense of the word. Dwelling on earth is living on earth, but it is also pondering and reflecting on earth, dwelling on it. Dwelling is taking the time to think about earth and our relationship to it. Adopting earth as our home is to dwell on what it means to live on earth and that earth is the only place where we can live. Adopting earth as home means to dwell on what it means to be earthlings, among so many other earthlings whose telling refusal is our constant companion.

Heidegger's notion of earth as *native ground* suggests a rootedness in history, but a rootedness that is always open to reinterpretation. The emphasis on *Heim*, in both the sense of *inhabits* and *home*, complicates any easy notion of inhabiting the earth as simply sharing the same finite surface. Inhabiting the earth is not just occupying the same planet. And sharing habitation is not just a result of the fact, as Kant points out, that the surface of the earth is limited. Although the unchosen character of our cohabitation of earth, suggested by Arendt, is resonant with Heidegger's notion of habitation, for Heidegger there are also the crucial elements of home and rootedness that constitute *the way in which we inhabit the earth*. Certainly, Heidegger insists on this attention to our mode of inhabiting the earth. We do not inhabit the earth merely as equally distributed across its surface, nor as merely cohabiting with others not of our choosing. While these are facts of our existence, it is also true that we call earth our home; our habitation is dwelling as belonging to earth as home. This is our way of being at home on the earth. For us, the earth is "native ground," *Heimatlicher Grund*, our home ground.

Our relationship to our history, and to our habitation, is essential to what it means to inhabit. For it is not just cohabitation that is unchosen. Habitation and cohabitation are both unchosen and chosen; they must be preserved through our attention to them, through which we can adopt them as our own, take them up anew, or rethink them. Habitation itself is unchosen in the sense that we find ourselves on earth in a particular milieu.[40] And, yet, while both habitation and cohabitation are indeed given, they are never merely given insofar as they are always and necessarily open to interpretation. Because we cannot control or master the ground of our own existence, we must interpret. Because we cannot know, we must interpret. As such, interpretation itself becomes a matter of ethics. We have an obligation to make what is given meaningful through interpretation. Making what is given meaningful is not mastering it or controlling it, but rather opening it onto different possibilities. This hermeneutic ethics, then, requires a fluidity that opens up rather than closes off possibilities. Hermeneutic ethics obligates us to open up rather than close off responses from earth and its inhabitants.

While the unchosen character of our cohabitation is metaphysically and epistemologically significant, it becomes ethically and politically significant when we take it up. Our ethical and political responsibility to question our relationship to the given is what makes us "human" insofar as being human is having a certain kind of relationship to earth and world. Moreover, because we belong to the earth as beings who can choose the unchosen character of our cohabitation, we have an obligation to do so. We have an obligation to question our relationship to the world, the earth, and their inhabitants because *not* doing so risks treating them as mere objects at our disposal or resources to be exploited. We have a responsibility, then, not to destroy each other or the earth. Not only surviving but also the thriving is at stake for human beings and other earthlings and the earth itself. Furthermore, we are dependent upon others with whom we share the earth, including other creatures. Not choosing them, not choosing to cohabit and share the earth with them, leads not only to their destruction but also to our own. And certainly it leads to the destruction of the biodiversity upon which our planet lives and thrives. The responsibility to choose our habitation, to take up our native ground, is an obligation to the earth, which entails obligation both to ourselves and to others. We have an obligation to make earth our unchosen chosen home, not in the sense of a comfortable familiar place, but rather fraught with the ambiguity of our relationship to home as *unheimlich*. We share the earth and the world with strangers, including other creatures, and their strange kinship is essential to the uncanny place we call home.

THE EARTH IN THE FOURFOLD

After the war, Heidegger's conceptions of earth and world take yet another turn with the introduction of the *fourfold (das Geviert)* and so too does his attitude toward home. Now Dasein finds itself embraced by the fourfold, dwelling in peace and freedom and the joy of returning home.[41] Heidegger introduces the concept of *das Geviert* in 1949, and it becomes the centerpiece of many essays from the early 1950s. The biggest change from earlier configurations is the addition of sky and the removal of world.[42] World is no longer one element among others in the fold of the four, but rather what results from their gathering. The world is presented as a gathering of the four, and things in the world are "gathering points" or clusters of "streaming relations" of the four that make up the world (Mitchell 2009, 208). The fourfold, then, can be seen as a further attempt to move away from both subjective and objective metaphysics and toward a profoundly relational phenomenology of experience of the worlding of the world or the happening of the event of Being. More literally, *Geviert* is an obscure German word for square or courtyard, also suggesting boundaries and limits, along with a sense of enclosed space. The *Geviert* is one's place as a dwelling place; the *Geviert* or fourfold is what makes a place a dwelling place, a place where we belong.[43] We can interpret this dwelling as a belonging through the fourfold by emphasizing both its essential relationality and its limits. In other words, we belong because we are profoundly relational and because we exist within the limits of the fourfold. And, yet, that we are bounded by the fourfold necessarily opens onto what is in excess of our existence, namely the fact that each element that makes up our experience is also beyond it.[44]

Within the fourfold, earth still appears as self-secluding and sheltering, but now in relation to the other elements of sky, mortals, and divinities. With the introduction of the fourfold, Heidegger no longer reserves *world* for Dasein, but rather focuses on the *relationality* of all beings, including Dasein, along with animals and stones.[45] Indeed, already in *Contributions*, Heidegger associates "world poor creatures" with "ones for whom the earth has always remained only something to be exploited," namely human beings (2012b [1936–1938], §252, 317). Here Dasein is something that must be achieved and not something merely given.[46] This achievement, then, is not in principle foreclosed for nonhuman animals.[47] The world is no longer the possession of Dasein alone, but rather the result of relationships between beings. Indeed, Heidegger is moving away from his notion of Dasein as world-forming and toward Dasein as protector of earth.

Even so, the notion that Dasein is an achievement still holds up the human way of being in the world as the goal, particularly insofar as Heidegger introduces the concept of *Dasein* to set this way of being in the world apart from the species human being. Thus, although Heidegger's view of the relationship between Dasein and nonhuman animals may have changed from his earlier infamous views in *The Fundamental Concepts of Metaphysics,* he still employs a comparative pedagogy in which all are measured against Dasein, which, for all his efforts to set it apart, is still human all too human.[48] For Heidegger, human beings are unique as a species because we can become Dasein. And, yet, what animal species is not unique in its way of being in the world? And who is to say that our way of being is superior? Certainly, we cannot be impartial judges of the relative worth of our way of being over that of other animals when we do not, and cannot, have access to their ways of being or their worlds. And while we can, and should, be held responsible for caring for the earth and its creatures, as Heidegger often reminds us, that responsibility does not, and should not, be taken to mean that we are in control of the earth or other creatures. Indeed, our agency with regard to our responsibility is part and parcel of our way of being in the world, not only as Dasein but also as human. For we are as much defined in terms of our species as any other, and so, too, every other species transcends its categorization as species, which, to be fair, is always only a human categorization after all.

Perhaps we can embrace Heidegger's deeply relational view of the earth and world without confining those relationships within hierarchies of relative worth. Rather, we can acknowledge that we are defined in terms of our relationships and that we are unique, which is to say both uniquely talented and uniquely limited, as is every species. Whether one's eyes are in front of one's head or both on one side of one's head is neither in one's control nor of value in itself. The worth of having the world before us rather than to our left is relative to our way of being in the world. For if we were halibut instead of human, it would put us at a decided disadvantage.

Within the fourfold, human beings are mortals and fundamentally relational beings. Our relationality entails limitations and possibilities, boundaries and promise. Mortals, divinities, earth, and sky coming together as four in one signal the dynamic relationality of all things.[49] Like the earth and every other element of the fourfold, the sky brings with it concealment and unconcealment, the unknown and the familiar, the uncanny and the homely (cf. Mitchell2009, 213, see also Mitchell 2014). In addition, the sky is both an inviting space but also the space of exposure, exposure to weather, the elements, the seasons, and the movements of the planets. Moreover, as exposure and unknown, the sky, like the earth, reminds us that we are not the masters of beings, or any element of the fourfold, even if for Heidegger our place within it, as mortals, is unique. As mortals, we are exposed to earth and sky; "the human sojourn" takes place between earth and sky, between birth and death" as exposure, which constitutes our fully relational way of being in the world (Heidegger 1957, 93).

Just as we are exposed to the elements and the sky, as mortals we are exposed to death.[50] And, yet, just as we do not control or possess the sky, we do not control or possess death. This dispossession is definitive of mortals.[51] We are marked by something outside ourselves, something beyond that we cannot possess, but nonetheless defines our way of being in the world. Our being toward death is a way of being exposed. To say we know what death is in truth is misleading and puts death back into the kind of metaphysical thinking that Heidegger attempts to avoid, especially in his later work.[52] As participants in the fourfold, mortals are no longer the "world-building Dasein" of Heidegger's earlier work. Rather, we have become dwellers who build only because we dwell (1977c, 350). Dwelling is also a way of being exposed to earth, sky, and mortality. Heidegger says that the verb *to dwell* "designates the manner in which man, upon the earth and beneath the sky, completes the *passage* from birth to death" (1957, 93; my emphasis). He emphasizes the passage "of the human sojourn *between* earth and sky, *between* birth and death" and calls this manifold "*between*" the *world* (93, my emphasis). The world is the time and place of the between in which human beings dwell and thus "the world is the house that mortals inhabit" (93). World is a way of dwelling on earth. As dwellers, we are thoroughly relational beings who occupy a between space and time that makes us utterly dependent upon other creatures around us and upon our environment, symbolized by earth and sky. Moreover, as dwellers, we have responsibilities to those creatures and the environment, to earth and sky.

Building as dwelling becomes associated with safeguarding not just Being but also earth. In fact, in this later essay, building is not so much about world as it is about earth. In his preface to the essay, David Krell describes what could be seen as another way in which the introduction of earth moves Heidegger further from the metaphysical tradition: "For Heidegger to dwell signifies the way 'we human beings *are* on the earth.' Man's Being rests in his capacity to cultivate and safeguard the earth, to protect it from thoughtless exploitation and to defend it against the calumnies of the metaphysical tradition" (Krell 1977, 345). Implicit in dwelling, then, is an ethics of the earth or an earthly ethics. We could say that as dwellers our mode of being in the world and on the earth is that of ethical creatures. Our way of being is cultivating and safeguarding the earth, what Heidegger also calls freeing the earth, in the sense of letting it go. Again, our ethical obligation to the earth appears as one of self-restraint. We have a responsibility to be responsive to the earth and its creatures. Otherwise we risk destroying it and ourselves in the process.

"Saving the earth," then, does not mean managing it, controlling it, penetrating it, knowing it, or revealing it. On the contrary, Heidegger describes saving the earth as freeing it: "To save properly means to set something free into its own essence. To save the earth is more than to exploit it or even wear it out. Saving the earth does not master the earth and does not subjugate it, which is merely one step from boundless spoliation" (1977c, 352). To save means not only to protect the earth from danger, but also to free the earth, to let the earth be without turning it into standing reserve for our own use or experimentation. Saving does not mean mastering, which Heidegger claims is just one step away from plundering the earth. Attempts to save the earth by using technology to predict and control it merely repeat the technological worldview that endangers it. Ultimately, for Heidegger, we must learn to dwell on earth insofar as to dwell means to save. In other terms, we could say that, resonant with Arendt's call to learn to love the world enough to take responsibility for it, dwelling is learning to love the earth enough to take responsibility for it. Furthermore, loving the world enough to take responsibility for it would mean taking responsibility for the ways in which our worlds and world-making impact the earth.

The human being has become a caregiver for the earth rather than a world builder. In fact, caring for the earth is radically different from making a world (see Heidegger 1977c, 349). Building as dwelling and dwelling as protecting, preserving, nurturing, and caring for the earth is not like making a world. The association between dwelling and earth alters our relation to world. No longer the world builder, human beings, as mortals, participate in the fourfold whose dynamic encounter gives rise to the world. As dwellers on the earth, and beneath the sky, we are obligated to protect and nurture the earth, which includes earthly creatures, and the sky, which includes elements in our environment. And yet, for Heidegger, the corner, or square—*das Geviert*—in which we find ourselves always limits this responsibility. We are bounded creatures, bounded by our response-ability and that of those around us and of our environment.

At this point, in order to begin to understand what our responsibility to save and protect the earth might entail, we must face perhaps the strangest element of the fourfold, the divinities.[53] Certainly, even more explicitly than any of the other three elements, the gods conjure something essentially beyond us.[54] Insofar as the gods are part of our experience and not some beings existing in Heaven, they remind us that there is always something outside our grasp. When Heidegger famously says, in a 1966 *Der Spiegel* interview, speaking of the reign of technology across the planet, "only a god can save us now," we could interpret him to mean that only something beyond ourselves, beyond technological standing reserve, can save us. Divinities, associated with the holy (*das Heilige*) and the whole (*das Heile*), cannot be assimilated into the technological mode of ordering the earth or the world; their participation in the fourfold prevents earth or world from becoming a "worldview" or "picture," which is even more relevant in this age of digital technologies and social media.[55]

Heidegger explicitly rejects religious doctrine or gods of theism; he often suggests that god(s) or *the absence of god*(s) can bring us into an encounter with the piety of thinking.⁵⁶ Perhaps even more than earth as self-seclusion, sky as unknown, and mortals as dispossessed by death, divinities are present in their absence from the fourfold. But, as Heidegger insists, this absence is not nothing: "The default of God and the divinities is absence. But absence is not nothing" (Heidegger 1971e [1951], 184). Rather, he describe it as the "liberation of man" from the unthinking of "the They" and "idle chatter" that makes up so much of our everyday lives.⁵⁷ This absence of God, or the absent presence of God, signals a beyond the human, which stands as a limit to our comprehension and mastery. Perhaps, more than any of the other three self-concealing elements of the fourfold, we are dispossessed by divinities and thereby disabused of any illusions of control. We are limited insofar as we are bounded by the fourfold.

And yet, for Heidegger, the limits or boundaries of the fourfold are not end points, but rather beginnings. For it is only through these limitations that relationships are possible, which suggests that things become what they are through the worlding of the world as limiting and bounding (cf. 1971c [1935–1936], 170). The limit is not just an end but also a beginning, an opening into the unconcealed and appearance of things in the world (208, cf. Heidegger 1977c [1951], 355). If the boundary or limit sets free, then saving the earth as freeing it involves recognizing our own limitations, not as an end point, but rather as a new beginning. Again, it is a matter of restraint, but, paradoxically, freeing restraint. Something can be free only within a boundary, only if it has limits. Indeed, it can be only by virtue of its limits. In this sense the end is another beginning. Moreover, limits are necessary for all beginnings and all creativity. As in the case of shadow and light, there is a reciprocal relation between the end or limit of one and the beginning of the other (cf. 2012a [1949], 129).

Acknowledging our belonging to the earth is a limit, to be sure. But, along with Heidegger, we can think of this limit or boundary in the sense of the Greek *peras*, which means *end*, as in *ends of the earth*, that comes from *peron*, which means *beyond* or *on the other side*. The limit as boundary is "that from which something begins its essential unfolding" (1977c [1951], 355). And meaning and creativity are possible by virtue of boundaries. Only because we are bounded do we have a world. Yet a world in which we exercise restraint and limitation of our own desire for mastery is essential to cherishing and protecting, preserving and caring for the earth. Realizing that as human beings we are bounded by both earth and world is also beginning to think of the ways in which our relationships to other creatures and to the earth itself always exceeds calculation, possession, or mastery and the ways in which we are open to change and surprise. Our relationship to earth comes through world; yet worlds can change as the earth "juts" through (note that *peron* is related to *pierce*). Acknowledging our limitations can be a beginning rather than an end. Heidegger describes a reciprocal relation between things, the fourfold, and the world that depends upon limits. Without limits, there is no thing and no world. But without excess and beyond, nothing would appear as thing or world, earth or sky, mortal or divinity. Relationships depend on both limits and excess, revealing and concealing, appearing and withdrawing, knowing and not knowing. Each element of the fourfold operates as both limit and excess that gathers things into the world and is gathered through things as they appear in the world.[58] Relationality is not only the essence of the fourfold but also of things in the world as they appear to us through the fourfold or as participating in it.[59] Relationality is the essence of human life insofar as human beings belong to the world; they are in the world and they participate in the fourfold as mortals (Heidegger 1971e [1951], 182). This participation both sets them apart and binds them to the other elements of the fourfold.

Through the participation of divinities in the fourfold, outpouring gestures are also gestures of gratitude. And thinking (*Denken*) becomes thanking (*Danken*). For example, one possible role of divinities in Heidegger's example of a bridge involves gratitude and thankfulness (1977c, 355). The same is true of Heidegger's analysis of the pouring and outpouring of thanks in his example of a jug in "The Thing." There he describes the fourfold as they are gathered in the jug in what he calls its "poured gift" (Heidegger 1971e [1951], 172). The jug brings together gifts from earth and sky to pour water or wine. But the jug can also be used for consecrated offerings to the gods, and, as such, "the gift of the outpouring as libation is the authentic gift" (172–173). For then the outpouring of the gift is not merely a filling and decanting, but rather an offering. In this case, its being as gift is explicit. Furthermore, it is a gift outside any economy of exchange; rather it is outpouring, again going beyond the self. In this giving as outpouring there is no expectation of reciprocation; there is no use-value or exploitation. The participation of divinities in the fourfold takes us beyond any possible reappropriation of our experience of the world into the calculable or technological ordering as standing reserve. There is no reserve, no stockpile, in this offering. Yet these divinities are not separate from earth, sky, and mortals; they all belong together in the worlding of the world. Heidegger maintains that when we try to comprehend each of the fourfold separately, when we try to order them according to calculable technological frameworks for thinking, we fall short. "The human will to explain just does not reach to the simpleness of the simple onefold of worlding" (180). This simpleness resonates with the pregivenness of the world and the earth such that they are always too close and therefore too faraway.

Thinking as thanking reaches toward the divine as it participates in our fourfold experience of things in the world. As such, it is also outside the economy of exchange and calculation. Heidegger describes thinking as thanking as a gift, not as recompense but as offering. He says that we owe thanks to what is most thought-provoking and that thinking about it is a form of giving thanks for it. This suggests that we owe thanks to what is question worthy and that questioning is a form of giving thanks. Implied in this debt is an ethical obligation to question and to think since we are endowed with a questioning nature. Yet, Heidegger insists, "This devoted thought is not something that we ourselves produce and bring along, to repay gift with gift. . . . Such thanks is not a recompense; but it remains an offering; and only by this offering do we allow that which properly gives food for thought to remain what it is in its essential nature" (1968 [1951–1952], 84).

Thinking as thanking, thinking as grateful attention, resists the technological mode of ordering the world into calculable reserves ready for use. This type of thinking is not thinking as representation; it resists seeing the world and everything in it as pictures. Whereas representational thinking is an attempt to master the world, grateful attention is a way of responding to the world. Whereas through representational thinking we set ourselves apart from the world as subjects viewing or using objects, through thoughtful offering we acknowledge our participation in the world as thoroughly relational beings (1977a [1938], 131, 132). As an alternative to the representational thinking of metaphysics, and the ordering and planning management of technological thinking, Heidegger proposes thinking as a mode of responding, responding with thoughtful questioning as grateful attention. First and foremost, this alternative mode does not position man as the subject or agent of thinking, nor does it allow him to take the position of mastery. Rather, human thought is a response and a listening, which may indeed be active and not merely passive, but never controlling or in control.

But, what does it mean to respond and listen rather than to master and control? In *What Is Called Thinking*, Heidegger explains, "The first step toward such vigilance is the step back from the thinking that merely represents—that is, explains—to the thinking that *responds and recalls*. The step back from one thinking to the other is no mere shift of attitude. . . . The step back takes up its residence in a *co-responding* which, appealed to in the world being by the world's being, answers within itself to that appeal" (Heidegger 1968 [1951–1952], 181–182, my emphasis). Listening requires vigilance and a shifting of perspective in relation to the world and others. Yet this shifting perspective is not something that we manipulate, but rather a responsive mode of being in relation to the earth and its creatures. What Heidegger calls the "step back" is responsiveness to the world and to the earth. And this responsiveness is an alternative to the technological mode of relating to world and earth as ordering, managing, and controlling. Thinking as responding is listening rather than mastering; it is letting be, rather than taking control or ordering (184). Thinking as responding is an alternative to both metaphysical thinking and technological thinking, both of which pit subject against object, man against world and earth.

This type of responsiveness requires not only acknowledging that we are deeply relational beings (dependent upon other people, other creatures, and the earth) but also attending to those others and our environment in ways that open up rather than close down the ability to respond. As Heidegger describes them, both the metaphysical and technological frameworks close down rather than open up possibilities for response; they allow only a fixed set of responses, if any true response at all. What Heidegger calls poetic thinking and dwelling, on the other hand, open up possibilities for response by exposing the ways in which every framework is an interpretation that could be figured otherwise. Poetic thinking allows an openness that shows the ways in which it is both limited and excessive. It is limited in that anything in order to be meaningful must be bounded; it is excessive insofar as meaning always exceeds those boundaries. It is precisely these limits and excess that both the metaphysical and technological frameworks deny.

More specifically, metaphysical and technological frameworks deny our participation in world and our existence on earth by imagining man standing over and against, or outside of, earth and world (figuratively, if not literally, as in the case of the Apollo astronauts). Both assume that through reason or technology man can dominate world and earth. And both thereby close off possibilities for responding otherwise or for true response. We could say that the technological worldview turns response into reaction and ensures that we can never truly respond. It turns us into machines, answering machines, that operate on autopilot rather than dwell, ponder, and respond. Since we are relational beings engaged with the world and living on the earth, rather than subjects standing over against objects, thinking is a matter of responding to our coinhabitants rather than representing objects. But opening up rather than closing off the possibilities of response requires vigilance. Caring for the world and protecting the earth require vigilantly safeguarding the ability to respond. We have a responsibility to enable response-ability.[60] The technological worldview cuts off response and disables the ability to respond.

Here again, response takes on an ethical tone insofar as it is better to listen and respond than it is to dominate and shut off. Moreover, enabling the response of others and the earth is a way of safeguarding them. And, as we have seen, dwelling means safeguarding, protecting, and nurturing. These "activities" require listening, attending, and letting be, rather than mastering or controlling. To safeguard and protect does not mean to lord over objects or horde them; it "can never be equated with the task of a guard who protects from burglars a treasure stored in a building" (1971e [1951], 184). Rather, it means to vigilantly attend to them by heeding "the call of being" (184). To safeguard and protect is not to guard or save things, but rather a way of being in the world and on the earth. Guardianship is responsiveness to the call of Being. It is a mode of listening through thinking and a mode of thinking as thanking. Certainly, this mode of relating to the world is radically different from the technological mode of controlling.

And yet, as Heidegger warns, "Everything here is the path of a responding that examines as it listens. Any path always risks going astray, leading astray"(1971e [1951], 186). There is always the danger that we will go astray. And the technological mode of ordering the world brings with it a greater risk of doing so. Andrew Mitchell describes the world of Heidegger's work from the 1950s as "a world that is always arriving, a fragile world shadowed by danger, but a danger that likewise allows us to belong to that world" (Mitchell 2012, viii). The world is fraught with danger, but also with saving powers, because the world is always open to reinterpretation.

GLOBAL TECHNOLOGY AS UPROOTING

In one of his most well-known essays, "The Question Concerning Technology" (1954), Heidegger analyzes both the danger and what he calls "the saving power" inherent in the technological mode of revealing the world as standing reserve. What he names the "extreme danger" is that man himself will become nothing but the orderer of the standing reserve, or, worse, "he comes to the point where he himself will have to be taken as standing-reserve" (1977b [1954], 332). And yet, at the same time, he exalts himself as the lord and master of the earth (332). It is not difficult to see this tendency today when advances in technoscience make many feel confident that one day we will "unlock" all the secrets not only of the human brain but also of the entire universe. For example, some proponents of genetic enhancements use precisely this language of mastery and control to describe the supposed moral obligation to enhance (e.g., see John Harris 2010).

Heidegger is clear that the danger is not technology itself, but rather the technological mode of ordering the world. While technology lends itself to the technological worldview, and while we can find technological causes for everything from disease and pollution to terrorism and the threat of nuclear war, Heidegger maintains that these effects are more fundamentally the outgrowth of our contemporary way of relating to both earth and world. Indeed, the logic of cause and effect, strictly speaking, is part and parcel of both the metaphysical and technological modes of relating to being. The alternative that Heidegger proposes is more difficult to understand. What is the saving power? Given Heidegger's criticisms of our illusion that we can master technology and thereby control earth and world, it seems that, contrary to what some have suggested, Heidegger cannot mean that the technology contains within it the power to save the earth, or the world for that matter, at least not in any straightforward way, as in the exhortation to "save the planet" from climate change (cf. Schalow 2006). Indeed, the planetary way of thinking is associated with the dangers of technology, so much so that we could say that *planet* works against both *earth* and *world*.

In terms of the "saving power," Heidegger is clear that it is not technology itself that can save humans or the earth. Just as the essence of technology is not anything technological, neither is its saving power. To the contrary, only by reflecting on the technological mode of revealing as one among others, attending to its dangers, and opening ourselves to an alternative relationship to the world can we hope to "save" the earth. Heidegger defines *to save* as "to fetch something home into its essence, in order to bring the essence for the first time into its proper appearing" (1977b [1954], 333). The saving power of technology, then, is precisely facing up to its danger, which is inherent in the essence of the technological mode of revealing as ordering, planning, and controlling. Thus attending to the essence of technology is neither something that we can achieve nor something that we can master. Rather it is a way of responding, of listening, that comes through pondering as questioning, which is not the same as answering or solving. Saving the earth is not a matter of solving a problem or calculating risks or pollution levels. For these technological approaches to saving the earth are part and parcel of the technological way of ordering that poses the danger in the first place. Indeed, the rhetoric of saving the earth is already loaded with presumptions of human control and domination. The fact that some people deny the existence of climate change because scientists have correlations rather than direct causalities, or because scientists cannot accurately predict or calculate the effects of man-made greenhouse gases as opposed to natural ones, only goes to show the dangers of committing to the technological mode of ordering and calculating as the sole mode of relating to the earth and the world. Heidegger's analysis suggests that as long as we try to use the instruments of technology to master or manage the earth, even by trying to save it, we are still caught up in the greatest danger. He says, "so long as we represent technology as an instrument, we remain transfixed in the will to master it," even if we see it as an instrument with which to save the planet (337).

The saving power involves thinking as endless questioning and openness to surprises, even wonder and awe, rather than trying to predict the outcome.[61] Heidegger describes man's role in the saving power as "keeping watch over the unconcealment—and with it, from the first, the concealment—of all essential unfolding on this earth" (337). In fact, given that technological enframing is dangerous because it positions itself as the sole way of revealing, vigilance toward concealment may be more important than guarding unconcealment (cf. 337). Perhaps for that reason Heidegger associates this concealment with *unfolding on earth* rather than in world. For, as we have seen, throughout the 1930s concealing is identified with earth while revealing is identified with world. With the introduction of the fourfold, concealing and revealing are operations of all four elements, including earth, as they gather things in their presence, but only by virtue of what is at the same time absent, as the coming into being, or *worlding*, of world. Perhaps for that reason, as well, in so many places throughout his work Heidegger talks of the importance of *roots* and *rootedness* and the dangers of *uprooting*. For, as Gadamer points out, in order to grow, living beings must be rooted on earth (in the case of plants, literally rooted *in* earth) and nourished by sunlight from the sky (1994, 106). Though we only see, or admire, the blooms above the ground, we must not forget that they are sustained by roots below that will shrivel and die if exposed to light. Some of Heidegger's favorites, the metaphors of roots and rootedness conjure earth every time he uses them. Heidegger insists that metaphysics is like a tree that has forgotten its roots in the earth, which nourishes and sustains it (1998a [1949]). Yet the relationship between the tree and the earth, between the roots and the soil, is never one of possession but rather one of intertwining.⁶² Rootedness is essential to all of life, not just biological life, but also the historical life of a people and the thoughtful life of Dasein. Heidegger even describes the saving power as *growing* and taking *root* (1977b [1954], 334).

In addition, he frequently talks of the modern technological age as *uprooting* man. One of the last places he does so is in the *Der Spiegel* interview when, after seeing images of earth from Lunar Orbiter I in 1966, he warns, "Everything is functioning. That is exactly what is so uncanny . . . that technology tears men loose from *the earth* and *uproots* them . . . I at any rate was frightened when I saw the pictures coming from the moon to the earth. We don't need any atom bomb. The *uprooting* of man has already taken place. The only thing we have left is purely technological relationships. This is no longer *the earth* on which man lives" (1981 [1966], 105–106, my emphasis). Heidegger suggests that global technology (symbolized by images of the earth from the moon) is threatening to *uproot* man from the earth. Furthermore, the planetary impetus of technology threatens to uproot even the earth itself. When warning of the "encounter between global technology and contemporary man," Heidegger uses the phrase *planetarisch bestimmten Technik*, which literally means "planetarily determined technics" rather than "global technology."⁶³ Again, Heidegger is opposing the planet and the earth. Technology aims its sites at ordering the entire planet through global communications and global markets. And, with the advent of colonies of settlers on Mars in the near future, if ventures such as

Mars One succeed, the earth becomes just one planet among others rather than uniquely our home.

This planetary way of thinking about our relationship to the earth is the result of the technological way of framing our experience. Planetary-determined technics are not just any form of technology. Nor are they merely technologies that attempt to "go global." Rather, we could say that the modern technological worldview that Heidegger warns about is always already global in its scope (see Lazier 2011, 611). Benjamin Lazier concludes, "The rise of the planetary in the modern imagination was synonymous for Heidegger with the demise of the earthly and the worldly, and these images from space only consolidated a process—a *globalization of the world picture*—already long in the making" (611, my emphasis). The technological way of approaching our relationship to both earth and world presupposes globalism, which threatens to destroy both.[64] Contemporary technology threatens the earth literally in terms of climate change, nuclear accidents or warfare, along with various forms of pollution and waste; conceptually, contemporary technology threatens the earth in terms of the concept that man can conquer the earth through technology, which not only disavows the ways in which earth resists or refuses to reveal itself to us but also positions the technological relationship to the earth as the best, if not only, relationship possible. Contemporary technology threatens the world by turning it into a worldview or world picture and thereby reducing our relationship with the world to one of subjects representing objects, again presupposing that we stand outside the world rather than participate in it.

Heidegger had already voiced similar concerns about the planetary when he compared Sputnik to metaphysics insofar as both threaten the complete technicalization of the planet (Heidegger 1971b [1950–1959], 58). Both Sputnik and the interplanetary information system homogenize and technicalize all communication and relationships on the entire planet, what Heidegger calls the complete "calculus of planetary calculation" (62). Insofar as Sputnik gave birth to the age of global communications, Heidegger's warning seems all the more apt today when the earth is surrounded by satellites relaying telecommunications around the globe. Traced through space, the orbital trajectories of these satellites looks like a second atmosphere completely enclosing the earth.[65] Sputnik was the beginning of a new information age of global telecommunications through which nearly instantaneous images from all over the globe connected only by the cacophony created by computer screens displaying the results from various web browsers. Images of Earth from space affect our view of earth, not only literally but also conceptually. Here again, the planetary and the global work against the earth as secluding, sheltering, and nourishing concealment.

Already, in 1935, Heidegger diagnoses the dangers of global technological thinking taking over the planet through our desire for speed and instant access to everything everywhere:

> Russia and America, seen metaphysically, are both the same: the same hopeless frenzy of unchained technology and of the *rootless* organization of the average man. When the farthest corner of the globe has been conquered technologically and can be exploited economically; when any incident you like, in any place you like, at any time you like, becomes accessible as fast as you like; when you can simultaneously "experience" an assassination attempt against a king in France and a symphony concert in Tokyo; when time is nothing but speed, instantaneity, and simultaneity, and time as history has vanished from all Dasein of all peoples.
> (2000C [1935], §29, 40)

This passage recalls Heidegger's emphasis on the notion of "a people" and its connection to earth. Earth in particular is associated with the native ground of "a people": and it is that native ground with which we must struggle as individuals and as "a people." This struggle is the constant interpretation and reinterpretation of what is given and what is pregiven, linked as they are to the earth. The world or worlds of "a people" are formed out of this struggle. Yet, as we have seen, how worlds are formed and human beings' participation in the activity of that forming changes in Heidegger's thought. No longer world-forming, but rather world-dwelling, we have a responsibility to take up both our past and future with the vigilance of a listening as responding.

The greatest danger is that we do not see the danger, and then technological ordering becomes the sole relationship we have with both earth and world: "The spiritual decline of *the earth* has progressed so far that peoples are in danger of losing their last spiritual strength, the strength that makes it possible even to see the decline and to appraise it as such" (Heidegger 2000c [1935], §29, 40). We are at risk of no longer being people or peoples when the "spiritual decline of the earth" reaches the point where there are no differences between peoples because technology has leveled all differences. When this happens, space and time appear as simultaneous and the past and the future become meaningless. The "spiritual strength" that Heidegger draws upon seems to be the struggle with the past and future that is definitive for individuals and for "a people." When that strife disappears, we risk the destruction of earth and the loss of world as the world "turns away." In this sense the global and planetary uproot not only man from earth and world but also world from earth.

Again foreshadowing the introduction of the fourfold, in *The Introduction to Metaphysics* Heidegger identifies this spiritual decline of the earth with "the flight of the gods, the destruction of the earth, the reduction of human beings to a mass, the preeminence of the mediocre" (2000c [1935], §34, 47). In addition, he warns that, because of this spiritual decline, "on the earth, all over it, a darkening of the world is happening" (§34, 47). When our only relationship to the world and the beings in it is one of ordering, or technological framing, then we risk the destruction of earth and the unworlding of world. If the world signals the relationality of beings, the between, then the technological worldview turns away from the space and time of the between and the relationality of experience and focuses on what can be calculated, quantified, managed, ordered, and controlled. Even the natural beauty of the earth becomes merely a landscape or recreation area. By turning relationships into things, the technological worldview threatens the destruction of earth and the unworlding of world.[66]

In light of the destruction of the earth and the unworlding of the world, man is left homeless, not in the sense of *Being and Time* as Dasein's ex-sisting, but rather by becoming part of the stockpile of reserves or a commodity to be ordered, consumed, even destroyed. Rather than belong to the earth or to the world, human beings treat themselves and other beings as belongings or property. The technological worldview distorts the sense of belonging and turns it into a form of possession such that the earth and everything on it appears as our possession. Alternatively, attention to the fourfold nature of the world recalls us to our dispossession as mortals who do not control even ourselves or our technology, let alone the earth, the sky, and the gods. Technological framing, on the other hand, perpetuates the illusions that man can master all things, including earth and sky, and that man can do away with divinity and spirituality altogether, which means doing away with anything beyond man.

When this happens, we even risk losing the ability to see the danger in this way of framing our relationship to ourselves and other beings, along with our relationship to both earth and world. We risk believing that ordering, planning, and managing are not only our primary responsibilities but also our only ways of relating to earth and world. Rather than listen to the earth's refusal, we refuse to believe that there is anything outside our control and mastery, that there is nothing we cannot penetrate. When the technological frame is our only framework, we revel in our own ability to reveal the truth of being while we disavow the ways in which beings resist us and wrest free of our grasp. Evidence of this takeover by the one-track mind-set of technological thinking is the veneration of science imagined as absolute truth unlocking the secrets of both the human brain and the universe and the simultaneous devaluation of poetic arts that not only resist any notion of absolute truth but also demand continuous interpretation and reinterpretation. In an economy evermore defined in terms of use-value and profit margins, these ongoing interpretative endeavors do not seem productive.

THE POSSIBLE SAVING POWER WITHIN GLOBALIZATION

At the end of "The Question Concerning Technology," Heidegger suggests that the poetic arts might be the *saving power* insofar as they illuminate a dialogue between humans and divinities (1977b [1954], 339). What Heidegger means by gods and divinities is open to interpretation. But their invocation indicates something beyond human control, something that we cannot know, that which remains a mystery. When he concludes, "questioning is the piety of thought," he suggests that the dynamic and continued activity of interpretation is essential to the "saving power" (341). Insofar as questioning takes us beyond ourselves and brings us closer to the mystery of divinity, perhaps "only a god can save us now" takes on another meaning resonant with hermeneutical thinking as questioning.[67] If not divine, at least questioning conjures the piety or reverence of thinking insofar as it takes us beyond the technological mode of ordering and toward a poetic mode of dwelling. Hölderlin's "poetically man dwells on the earth" becomes associated with Heidegger's suggestion that questioningly "man dwells on the earth" (340). For poetic thinking is a thinking that continually opens itself to questioning, to interpretation and reinterpretation. And dwelling as preserving through thinking as questioning is an alternative to the technological way of ordering experience. Describing the way in which the fourfold "orders" a farmhouse in the Black Forest, Heidegger suggests that, unlike technological ordering, this alternative way of ordering takes us beyond ourselves and calls on us to respond to the environment (1977c [1951], 361). Dwelling is a response to what presents itself, what is all around us in our world and on the earth (cf. 1977b [1954], 323). Moreover, it is a mode of attending to what does not present itself to us, what is beyond our experience and yet conjured by it. Dwelling is a way of being at home with the earth's refusal.

In "Building Dwelling Thinking" Heidegger associates man's homelessness with a lack of questioning, specifically questioning what it means to dwell (1977c [1951], 363). Dwelling is our way of being on the earth, which is to say our way of being is to question what it means to dwell on the earth. And to think for the sake of dwelling is to question what it means to dwell, not just anywhere, but on the earth and as mortals. It is only when we believe that we have answered the question of dwelling that we no longer consider it as "plight" and, as a result, we become consumed by the technological relationship to earth as a resource that exists solely for our use and exploitation. When this happens, our relationship to the world as a world of meaning through questioning becomes unworlded as questioning disappears. Dwelling must remain our plight, both in the sense of plight as situation and in the sense of plight as pledge.[68] As noted earlier, the German word *Not* which is translated as "plight" connotes urgency, emergency, and necessity, which are not present in the English translation. Dwelling as plight and the plight of dwelling again reminds us of the urgency of our ethical obligation to the earth. We are dwellers on earth; dwelling is our plight. And yet we must make dwelling our plight, our pledge or vow. The task of taking it up is an urgent one. We must vow to dwell as nurturing and protecting the earth and its inhabitants.

When we cease questioning, we cease dwelling. Thus we have an ethical responsibility to continue questioning. Heidegger's writings on technology insist on the urgency and vigilance required for this questioning as dwelling and dwelling as questioning. For, within the technologically mediated world, questioning and openness to questioning is an endangered species that risks extinction. And, if and when that happens, then the destruction of earth and the annihilation of world may take hold. This questioning is not the technological curiosity that demands satisfaction by insisting on penetrating into every being and forcing everything to reveal itself. Rather, questioning as poetic dwelling ponders our place amongst beings in the world and on the earth not as subjects facing objects or owners managing resources, but rather as restraining ourselves so that beings can "shine" (cf. Heidegger 1977b [1954], 341).[69] The illumination of beings does not come from us. Rather, it is given to us as a gift. In this regard, thinking as questioning is a form of gratitude for that gift. And gratitude is a way of respecting the refusal of earth and its inhabitants to reveal everything about themselves. For, in the mystery of their concealment lies what Heidegger calls the "awesome" that engenders "astonishment" (1971c [1935–1936], 200–201).[70] The gift is a surprise, which is possible only because we do not own the earth, but rather belong to it. And we can belong to it only because we are not gods, but rather mortals. We belong to the earth because we are creatures living amongst, and cohabiting with, countless other earthlings.

Poetic dwelling is openness to the beyond human—the other than human—that which withdraws from view, what cannot be mastered, what remains concealed.[71] Poetic dwelling is attending to the earth's refusal of complete revelation and light; it is respecting the need for darkness and submersion for rootedness. Poetic dwelling is also restraint, the holding back and stepping back of our own will to power. As an alternative to the lure of the single-mindedness of technological framing, poetic dwelling opens up the possibility of mindfulness and vigilant watchfulness as listening to, and for, the otherness of being, that is to say the ambiguity and the uncanniness. Only when we dwell poetically on the earth can we see beyond ourselves and experience the earth as something worth saving, not through our technological prowess but rather through restraint and sacrifice in the sense of offering to what is beyond human, including the rest of our awe-inspiring biosphere.

Invoking the stars in the heavens, poetically, Heidegger suggests the difficult path of restraint in the face of the lure of technology, "The irresistibility of ordering and the restraint of the saving power draw past each other like the paths of two stars in the course of the heavens. But precisely this, their passing by, is the hidden side of their nearness. . . . Ambiguousness is constellation, the stellar course of the mystery" (1977b [1954], 338). Without ambiguity and mystery, there is nothing out of the ordinary, nothing awesome or astonishing. The danger and the saving power are so near to each other that their ambiguousness is their constellation. Like the stars in the heavens, ordering and restraint are in the same orbit.

The Earth seen from space provoked, and continues to provoke, both these tendencies, the desire to order and control, on the one hand, and the desire to save through self-restraint, on the other. As we have seen, with the first pictures of Earth from space, the "irresistibility of ordering" and a certain discourse of something like "the restraint of the saving power" not only drew past each other like orbiting planets but also collided head on in the changing attitudes toward earth initiated by seeing it "whole" for the first time. The One-World movement captured the sentiment of technological imperialism of the planet, while the Whole Earth movement signaled the possibility of a discourse of saving the planet, if not quite the saving power that Heidegger imagines. The images that scared Heidegger from Lunar Orbiter I were black and white and show a deserted and desolate moon looming in the foreground, completely overshadowing a tiny slice of earth in the background, so small in some photos that you can barely see it. The later stunning color photographs taken on the Apollo missions continue to captivate with their uncanny beauty and loneliness. And if these photographs of Earth from space brought about the "globalization of the world picture," they also conjure meditations on our uncanny relationship to the earth that might return earth to the globe (cf. Lazier 2011, 623). Indeed, this may be the "saving power" of the "globalization of the world picture."

The technological prowess of the Apollo missions was not their only enduring legacy. For, what were planned as moon missions ended up turning our attention back toward Earth. Seeing the earth "whole" seemed to produce the realization, as if for the first time, that human beings are all of the same kind; and moreover, share the same earthbound home. Decades ago, when the first images of Earth from space were transmitted, they were met with the sentiment that all men are brothers, that all humans are united by our common planetary home. And because we may be alone in the universe—or even more so if we are not—in spite of our difference, we are humankind. As we have seen, the Apollo missions were touted as goodwill missions for all mankind. And yet, although this rhetoric belies the nuclear arms race of the cold war that inspired the missions, and while those photographs couldn't show us the earth "as it truly is," the images of Earth floating in space and the astronauts circling the globe beyond Earth's atmosphere did give many on earth a sense of a common home, our planet.

Today these images are ubiquitous. And yet, with the climate crisis looming, again we see the earth as our fragile home and again we hear calls to "save the planet." Again we imagine leaving the Earth, escaping it even. Can the environmental crisis bring us together as "a people" with a common home and a common purpose? Perhaps the environmental crisis includes within it the saving power of uniting nations and peoples of earth with the purpose of reversing the damaging effects of pollution and saving our earthly home. The potential saving power within the environmental crisis may be the acknowledgment that although we do not choose to live on earth, we can elect to make it our home. This means not only reenvisioning what we call home, but reenvisioning with whom we share it. Can the earth serve as our native ground in the Heideggerian sense of both given and chosen? We are all earthlings—that much is given. But, what would it mean to dwell on the earth *as earthlings*? It would mean that we dwell on the earth as earthlings in a way that responds to the earth and our coinhabitants by questioning and pondering what it means to be earthlings. It would mean reflecting on how considering ourselves earthlings among so many others, all bound to the same planetary home, changes the ways in which we think about those others and our obligations to them. It means loving the earth enough to take responsibility for it as earthlings dependent upon each other and upon the planet.

Certainly Heidegger resists globalization in favor of native grounds of *peoples* living on earth. Yet, perhaps the "saving power" of contemporary global technologies is the ways in which they unite people and give us common goals, and maybe these goals are shared not just by human beings but also, at least implicitly, by other species on earth insofar as each wants to survive and thrive. Furthermore, perhaps climate change is the issue—the most pressing issue of our era—that gives us a common native ground in the earth itself. *The earth of the Earth*, so to speak. In fact, insofar as all earthlings are affected by climate change, it could be the issue that pushes us beyond ourselves to consider other creatures with whom we share the earth. Perhaps the ambiguous "saving power" of global technology, the underside of its danger, is the way in which it brings us together as creatures who belong to the earth that is to say, as earthlings.

Indeed, the photographs of Earth from space continue to produce the uncanny sensation of realizing that we are essentially earthbound beings and that somewhere on that planet each one of us lives. In *Pale Blue Dot* Carl Sagan reminds us of the uncanny fact that "on it everyone you love, everyone you know, everyone you ever heard of, every human being who ever was, lived out their lives . . . every saint and sinner in the history of our species lived there—on a mote of dust suspended in a sunbeam" (Sagan 1994, 6). But shouldn't we say every animal you love, every creature you know, every living being you have ever heard of lives there, which is to say, here on our shared home. For seeing ourselves as earthlings takes us beyond the human and toward the biosphere, full of life. The fate of every living being is tied to this planet that we call home, with all of its ambiguity and mystery.

5

THE WORLD IS NOT ENOUGH

Derrida

Although the earth does not occupy a central place in Jacques Derrida's writings, his last seminar is all about the world. More exactly, it is about the destruction of the world that results from the death of each singular living being and the disappearance of the world necessary for ethics. Derrida begins his exploration of the receding world by following in the footsteps of Robinson Crusoe as he retraces them over and again on the desert island. Derrida takes up the question of what is an island, and his analysis of Robinson Crusoe revolves around the ways in which the island both is and is not deserted, as well as Robinson Crusoe's ambivalent relation to the island, his attraction and repulsion. Although Derrida is not thinking of "this island earth," imagining the Earth as an island, isolated and alone, floating against the vast sea of space, as the astronauts did, conjures the ambivalent reactions he attributes to all islands. In *The Beast and the Sovereign Volume II*, Derrida asks, "Why does one love islands? Why does one not love islands? Why do some people love islands while others do not love islands, some people dreaming of them, seeking them out, inhabiting them, taking refuge on them, and others avoiding them, even fleeing them instead of taking refuge on them?" (Derrida 2011, 64). He continues this line of question, asking what is it that one flees or seeks when one escapes from, or takes refuge on, an island. He proposes that we find these seemingly contradictory desires in the same person, even in "the same desire" (69). We are both attracted to and repulsed by islands, caught between "insularophilia and insularophobia." But what is an island that it provokes this "double contradictory movement of attraction and allergy?" (69). While Derrida doesn't answer these questions head-on, he does suggest that the logic of *autoimmunity* is operating in figures of islands as isolated or insular and circular or round, floating alone as solitary and unique. Following Heidegger's discussion of solitude, Derrida finds in islands the double senses of being alone as being lonely and isolated, on the one hand, and being unique and singular, on the other. Certainly, these two senses of "alone" are operative in reactions to seeing the photographs of Earth from space. In the case of Robinson Crusoe, this autoimmunity shows up in his desire to leave his family, particularly to get away from his father's authority, and the stifling conventions of his home country, only to attempt to recreate that authority and those conventions on the island. Like a wheel turning in on itself, Robinson Crusoe becomes the very thing he is trying to escape. In conjunction with his analysis of Robinson Crusoe, Derrida associates autoimmunity with the wheel turning on its axis. The wheel is not just any circle because, like the Earth, it turns around its own axis. Like the Earth, the wheel is not stationary. And, like the wheel in Derrida's account, the Earth itself, spinning on its axis, triggers an autoimmune response insofar as images of the Earth as an island, beautiful and blue, floating alone in the darkness of space, are both threatening and reassuring. Circular shapes and circular movements such as Robinson Crusoe's island and his circumnavigations of it, along with his wheel, appear as metaphors for the return to self that his shipwreck instigates, which signal further parallels with the metaphor of Earth as island.

Recall that images of Earth lead us to see the Earth both as our amazingly singular home and as a tiny insignificant pea barely visible from space. The view of Earth from space makes us both want to protect our vulnerable and fragile planet and to escape from this insignificant speck in the universe. It makes us feel simultaneously special and inconsequential. This is the double sense of the loneliness of Earth; it is all alone in the universe and yet unique, a loneliness that resonates with Heidegger's notion of solitude as being alone as in without others and being alone as in without equal. This double desire is the ambivalent desire that Derrida associates with islands, both wanting to take refuge and wanting to flee.

Recall too that the media reports immediately following the Apollo missions' images of Earth from space are full of rhetoric of "returning home," returning to self, of man's finding himself, of man becoming his "true self," and so on. The iconic images of Earth from space became metaphors for man's homecoming, united on one planet as "brothers in eternal cold," and symbols for man overcoming his own worst tendencies in order to master himself. In the words of *Time,* the hope of Apollo 8's mission is that "as man has conquered the seas, the air, and other natural obstacles, he has also at each stage, in a small way, conquered part of himself. Therein lies the hope and the ultimate promise of his latest conquest" (*Time* 1969, 17). Or, in the more optimistic tones of Archibald MacLeish, "man may at last become himself" (MacLeish 1968).

Given that the Apollo missions that transmitted the first images of Earth from space were products of the cold war, MacLeish's remark about brothers in eternal cold might take on a different hue. Indeed, as we have seen the space program was driven by the conflicting rhetoric of nationalism and cosmopolitanism. The Apollo missions were both attempts at "winning" the cold war and putting "America First," and yet they were sold as missions for all mankind, intended to unite human beings across the globe. One small step for man, one giant leap for mankind. The tension between nationalism and cosmopolitanism has been a leitmotif in our discussions of both Kant and Arendt. Derrida's intervention in this debate is well known. For, his discussion of cosmopolitanism revolves around the notion of hospitality, which he extends beyond the Kantian conditional hospitality, insisting on unconditional hospitality and the right of refuge. And yet he also contends that cosmopolitanism, or what he calls *cosmopolitics*, necessarily puts into tension unconditional and conditional hospitality (Derrida 2001, 4, 5). For Derrida, there is ambivalence at the heart of hospitality.

We have seen the ambivalence inherent in hospitality—ambivalence between host and hostage, between hospitality and hostility—as it is manifest in the rhetoric around the Apollo moon missions and the first photographs of Earth from space. On the one hand, these missions were part of a military operation to secure the earth from hostile enemies; on the other hand, they were seen as benefiting all humankind, uniting mankind as "brothers" who share a common world and common goals. In terms of Derrida's analysis of hospitality, it is clear that implicit in the welcoming gesture of the cosmopolitan rhetoric of uniting all of humankind is the affirmation of the technological superiority of the United States and its concern to dominate not only the earth but also space. As the "victor" in the cold war,

America was in the position to extend hospitality to the rest of the world. As Derrida points out, hospitality is not only about generosity, but also always about control and mastery of what one takes to be one's own home (17). The ambivalence surrounding the Apollo missions and reactions to them resonates with imaging the earth as an island, as NASA did when it published the glossy photo-filled book *This Island Earth* shortly after Apollo 17's mission to the moon.

The ambivalence of hospitality resonates with the ambivalence of islands, whether it is "this island Earth" or Robinson Crusoe's island. In the case of Robinson Crusoe, he longs for human companionship more than anything else on his deserted island. And yet, when he discovers the footprint in the sand, he is terrified and longs for solitude. He both wants, and fears, company on his island. The supposedly "deserted" nature of his island "home" is threatened by the "savages" he witnesses on the beach. He is both fascinated and repulsed by these strangers and intends to kill or enslave them. He wants to flee the island, yet he returns to the island, which represents both the terror of solitude and the blissful paradise of solitude. Robinson Crusoe's ambivalent relationship to his island is akin to the ambivalence witnessed after the Apollo missions, namely the terror of our solitude on earth and the absolute uniqueness of our earthly paradise in the vast emptiness of space.

As we have seen, Kant identifies the earth with an inhospitable hospitality that echoes man's asocial sociability and points to this same ambivalence. For Kant, the goal of perpetual peace through universal hospitality is to overcome this ambivalence for the sake of equilibrium that brings peace, even if this goal is only a regulative ideal and can never be achieved. Derrida goes further when he argues that hospitality is more than a Kantian regulative ideal precisely because there is an internal contradiction inherent within the very notion of hospitality itself (Derrida 2000, 149). He points to this contradiction in this question: "In giving a right, if I can put it like that, to unconditional hospitality, how can one give place to a determined, limitable, and delimitable—in a word, to a calculable—right or law?" (147–149). In other words, the principle grounding all conditional hospitality, namely unconditional hospitality, is at odds with its practice. For, what makes hospitality unconditional not only makes hostility possible, but also inevitable insofar as ultimately there is no calculus with which we determine how to distinguish one from the other. The threat to unconditional hospitality does not come from outside, but rather from inside. Hospitality operates according to the autoimmune logic distinctive of all appeals to the self or sovereignty. In other words, if, or insofar as, hospitality is granted by one to an other, its unconditionality is already comprised. Indeed, the very terms *self* and *other* are problematic if our goal is unconditional hospitality; but these terms are required by our notion of hospitality insofar as we imagine that someone has the power to extend hospitality to another. And yet this very power acts as a condition that prevents hospitality from being unconditional.

Thus we must be vigilant in watching for the threat to unconditional hospitality from within our attempts at hospitality themselves. We must be watchful for the ways in which extending hospitality may also extend hostility or extending hospitality to one may exclude another. If we cannot ground unconditional hospitality on a universal principle that does not also always undermine itself, then we have only the groundless ground that is, in our terms, the earth itself. We must extend hospitality because we must coexist. And we must coexist because we all—all living beings—have a unique and singular bond to the earth. The earth is our home. And yet, as Derrida points out, "home" is precisely what is at stake in hospitality. The "problem of hospitality," he says, "is always about answering for a dwelling place, for one's identity, one's space, one's limits, for the *ethos* as abode, habitation, house, hearth, family, home" (Derrida 2000, 149–151).

Because hospitality is associated with home as *ethos*, "ethics is hospitality" (Derrida 2001, 17). And yet ethics as hospitality or hospitality as ethics points to the tension between ethics, so understood, and politics. Indeed, one way of articulating the contradiction internal to the notion of hospitality is in terms of the conflict between ethics and politics. Ethics of hospitality demands that we welcome every singular being in its singularity—already the words *we* and *its* belie the impossibility of such a demand—and politics, even a politics of hospitality, demands that we develop a universal principle of hospitality that applies to all, effacing the singularity required by ethics. In other words, ethics demands consideration of the singularity of each unique being, while politics requires universal rules and principles that apply equally to all. Hospitality, then, must be synonymous with ethics and yet exceed politics and morals, if by morals we mean principles that we can apply to action.

In Arendt's terms, we might say that *the right to have rights* can never be grounded in any practical right; furthermore, the right to have rights is at odds with any practical policy that grants rights insofar as "rights" are themselves always limited, granted by some to some and never by all to all.[1] Kant recognizes this problem in his complicated justification for private property on the common possession of the surface of the earth and the assumed social contract that follows from it. Indeed, all particular rights and the legal prescriptions that circumscribe them cannot touch what grounds the right to have rights, which is the value of each one within the plurality of human beings. Moreover, if rights are based on belonging, whether to a nation-state or an international community, ultimately they must be grounded on the earth. Insofar as all human and nonhuman living beings must live on earth, they belong to the earth, and any right to have rights must be grounded on the earth. We could say, then, that both unconditional hospitality and the right to have rights are grounded on what, in Heideggerian terms, we might call the *groundless ground* of the earth.[2] But, once we make this move, we must also extend the Arendtian right to have rights to all living beings, and perhaps beyond, insofar as each one is valuable, even essential, to the plurality or biodiversity of our cohabitation on earth.[3] As Derrida reminds us, hospitality is also about home and belonging. Here we push Arendt's notion of the right to have rights toward Derrida's extension of Kant's notion of hospitality when we ask: To whom does the earth belong, if all living beings necessarily and singularly belong to it? Who's home is it, if all of us by necessity live here?

The ambivalence inherent in hospitality speaks to ambivalence in the concept of *home*, which is complicated, to say the least, when we consider earth as our home. If "ethics is hospitality," then ethics is about home, not only because home and ethics share a common Greek root, *ethos*, but also, and moreover, because struggles over home and hospitality are at the heart of our relationship to others, particularly when considering the earth as home and our relationship to nonhuman beings. "Insofar as it has to do with ethos," says Derrida, "that is, the residence, one's home, the familiar place of dwelling, inasmuch as it is a manner of being there, the manner in which we relate to ourselves and to others, to others as our own or as foreigners, ethics is hospitality, ethics is so thoroughly coextensive with the experience of hospitality" (2001, 16–17). And yet Derrida argues that the history of hospitality bears out its ambivalence insofar as it is a history of violence and hostility. Being at home with the other, even with the other within oneself, raises the specter of the uncanny, which can provoke violence. Or, in the best of cases, perhaps poetic affirmation of each one can become the love of the world and of the earth through which we value each life, not only insofar as it contributes to plurality and biodiversity, which, as Arendt puts it, is the law of the earth, but also insofar as the singularity of each makes up a world, perhaps even the world. Certainly, when we consider earth as home, we encounter the uncanny at every turn, whether it is the uncanny strangeness of other animal creatures or the uncanny strangeness of the earth itself as seen from space, and these experiences fill us with ambivalent desires. Does our fascination with the other tip over into abjection and lead us to violence? Or, can it lead us to appreciate difference and what Heidegger calls the *awesome* or *mystery* of this uncanny encounter? Perhaps through poetic affirmation we can point to the groundless ground of hospitality and the right to have rights, which necessarily move us out of politics and morals, with their rules and laws, and toward an ethics of earth, an ethics that can guide practical morality and political action.

Taking us further than Kant, Arendt, or Heidegger, Derrida embraces a cosmopolitanism based on the radical singularity of each living being. With Derrida, we move from Kant's universalism through Arendt's pluralism to the absolute singularity of each as not only *a* world, but also *the* world. Derrida takes up the question "what is the world?" in the context of reading Daniel Defoe's *Robinson Crusoe* together with Martin Heidegger's *Fundamental Concepts of Metaphysics: World, Finitude, Solitude*.[4] Following in Heidegger's footsteps, Derrida explores the connections between *world, finitude,* and *solitude* through the vehicle of Robinson Crusoe's isolation on his desert island where he lived for years as if he were the only man on earth—that is, until he found that fateful footprint, human, all too human. Challenging both Heidegger's thesis that animals are poor in world while humans are worldbuilding, along with Robinson Crusoe's sovereign reign over his island home and all its beastly—or, more accurately, *beastie*—inhabitants, Derrida suggests that each singular living being inhabits its own solitary world, its own desert island. Even while staking this claim, however, Derrida attacks the sovereignty granted to human beings alone in both Heidegger's seminar and *Robinson Crusoe* by making the double movement familiar from his earlier work. For example, in *The Animal That Therefore I Am*, on the one hand, animals may possess reason or language or the ability to respond, or whatever other characteristic we usually reserve for humans alone, and, on the other hand, we cannot be certain that human beings possess these characteristics or abilities that supposedly distinguish us so clearly from other animals.[5] In *The Beast and the Sovereign, Volume II*, Derrida makes a similar double move when he claims both, on the one hand, that animals share our world and may be world-building and, on the other, that we cannot be certain that human beings share a world or are world-building (at least not in Heidegger's sense as set apart from animals). As isolated as Robinson Crusoe is on his desert island, it turns out that his island is not as deserted as he thinks. He is alone, but still radically dependent on others. This chapter traces the ethical implications of Derrida's seemingly contradictory claims that we both share a world and that each singular being, like an island, is a world unto itself. For Derrida, ethics begins where the world ends. This is to say, that ethics begins outside of any set of rules or principles, any common language or grammar, any shared culture or traditions. It begins in the ethical obligation to a singular other. This obligation to the singularity of the other is the force of the ethical bind. Yet this bind can only be lived and felt within some shared world. Like Heidegger's earth that juts through the world, the ethical bind between singular beings juts through our shared world and shatters it and yet thereby shelters it.

DEATH AS SUCH

Derrida's investigation of world in *The Beast and the Sovereign, Volume II* is motivated by Heidegger's claim that the stone is worldless (*Weltlos*), the animal is poor in world (*Weltarm*), and human beings are world-building (*Weltbildend*). As we know, Heidegger's comparative analysis of animals and humans revolves around the question of having or not having world. For Heidegger, humans clearly have it, while animals do not, quite. More specifically, animals do have a relation to the world, but not to the world *as such*. Humans, on the other hand, have relations to both the world and the world as such. Derrida challenges Heidegger's notion that Dasein has access to the world *as such* any more than animals do. He asks, "Have you ever come across the world as such?" (Derrida 2011, 268). As he does elsewhere, Derrida questions Heidegger's confidence both that animals do not have access to the *as such* and that Dasein does.[6] Although there are several texts in which Derrida criticizes Heidegger's position on animals, *The Beast and the Sovereign, Volume II* is not only the most sustained analysis of *The Fundamental Concepts of Metaphysics* but also goes beyond others in terms of the focus on the world. It is clear from Heidegger's comparative analysis in the seminar that he is trying to answer the questions What is world? What is finitude? What is solitude? In other words, Heidegger's goal is not to write a treatise on animals, but rather to determine Dasein's relationship to the world. Animals are pedagogical tools that he uses along the way (cf. Oliver 2010). Appropriately, then, Derrida's critical engagement with Heidegger's seminar revolves around the world as much as it does around animals.

Noteworthy, however, in Derrida's analysis of Heidegger's *Fundamental Concepts* is how quickly his discussion of world and the animal's relation to world leads to a discussion of death. Indeed, it is striking that, in over three hundred pages, Heidegger mentions the animal's relation to death only once, at the end of chapter 5, where he makes his famous pronouncement: "Because captivation belongs to the essence of the animal, the animal cannot die in the sense in which dying is ascribed to human beings but can only come to an end" (Heidegger 1995 [1929–1930], 267).[7] While it is true that this pronouncement echoes similar claims that Heidegger makes in other texts, in *Fundamental Concepts* he is more concerned with distinguishing animal behavior from human comportment and the way in which the animal's "disinhibiting ring," as he calls it, prevents it from comporting itself and in turn from accessing the world *as such*.[8] Given that Heidegger barely mentions death in *Fundamental Concepts*, how, then, does it come to dominate Derrida's discussion of this text in *The Beast and the Sovereign*? Certainly, we might think that Derrida's focus on death is warranted by the centrality of the notion of being-toward-death in *Being and Time*. This might justify why Derrida links Heidegger's long meditation on various moods of philosophical inquiry—including boredom, nostalgia, and melancholy—to death. Most obviously, the finitude of Heidegger's subtitle for his seminar—*World, Finitude, Solitude*—implies death. Yet, as we will see, Derrida does not make these obvious moves. Rather, he moves from world to death, not through an analysis of Heidegger on finitude, but by a certain substitution or displacement of a fragment from Novalis central to Heidegger's meditation on world and philosophy for a fragment from Paul Celan, which seems to haunt Derrida throughout this text and others.[9]

Heidegger quotes Novalis, "Philosophy is really homesickness, an urge [*Trieb*] to be at home everywhere" (1995 [1929–1930], 5). Heidegger interprets this fragment as meaning that philosophy does not want to be at home here or there or even in every place, but rather in the world as a whole:

"This is where we are driven in our homesickness: to being as a whole. Our very being is this restlessness. We have somehow always already departed toward this whole" (5–6). Derrida translates and comments on this passage: "It is toward this [he has just named the world: what is the world? Reply:], toward this (*Dahin*), toward Being as a whole (*zum Sein im Ganzen*)—it is that toward which we are driven (*getrieben*) in our nostalgia.... The nostalgic push or drive is what, basically, far from pushing us toward this or that, Ithaca or England, is what pushes us toward everything, toward the world as entirety" (2011, 101, brackets in the original). Derrida concludes this interlude on Heidegger's nostalgia by relating it to the fragment from Celan: "\ *Die Welt ist fort, ich muß dich tragen* (The world is far away, I must carry you; 104). Novalis's philosophical homesickness has become Celan's distant world that obligates carrying the other.

As we will see, this line of poetry becomes the axis around which Derrida's analysis of world, death, and ethics revolves. Through this poetic fragment Heidegger's world as the totality of beings becomes Derrida's world far away. And finitude becomes not just our relationship to our own mortality, but our being toward the death of the other (even the other in ourselves). Solitude, which for Heidegger makes us both alone and unique, for Derrida becomes the singularity of each living being insofar as we both do and do not share the world. Through the Celan fragment, Derrida translates Heidegger's *World, Finitude, Solitude* into absence of world, death as always the death of the other, and the singular ethical responsibility that both separates us from the world and binds us to it.[10] Repeatedly returning to the Celan fragment, Derrida's seminar suggests that ethics begins where the world ends and vice versa; the end of the world is the beginning of ethics. My obligation to you, to the other, starts where the world ends, when it is faraway and gone. And, when "the world is gone, I must carry you." *Die welt ist fort, ich muß dich tragen*. Heidegger's World, Finitude, Solitude becomes Derrida's Worldlessness, Death, Responsibility.

THE POETIC AXIS OF ETHICS

The first mention of this line from Celan's poem ("Vast Glowing Vault") in *The Beast and the Sovereign, Volume II* is in session 1 when Derrida lays out his main theses for the course in the form of three questions/sentences: 1.

What is an island? 2. The beasts are not alone. 3. What do beasts and men have in common? (2011, 3–8). Then, in answer to the third question, Derrida sets out three responses; and it is here that he appeals to the Celan poem. Derrida outlines three possible true answers to the question of what beasts and men have in common: 1. animals and humans inhabit the same world, which he qualifies as the same "objective" world, even if they do not have the same experience of those objects; 2. animals and humans do not inhabit the same world, "for the human world will never be purely and simply identical to the world of animals"; and 3. no individual animal or individual human inhabits the same world as any other (9). The third and most radical claim is the one that Derrida elaborates at length, calling the poetry of Celan as his witness.[11]

Describing the implications of this thesis, Derrida says, "between my world and any other world there is first the space and the time of an infinite difference, an interruption that is incommensurable with all attempts to make a passage, a bridge, an isthmus, all attempts at communication, translation, trope, and transfer that the desire for world or the want of a world, the being wanting a world will try to pose, impose, propose, stabilize. There is no world, there are only islands" (9). This radical hypothesis—that there is no common world and that each living being is separated from every other like an island—is immediately followed by the first appeal to Celan's poem.[12] Derrida suggests his claim that there is no world, only islands, is an interpretation of the Celan fragment. While it may be more obvious why he would say that this line of poetry can be interpreted to mean that there is no world—the world is gone—it is less clear why it can be interpreted to mean that every living being is an island, especially given that the poem still includes the pronouns *I* and *you*. As we will see, this performative paradox, which announces the end of the world, even while maintaining *I* and *you* and a relationship between the two, becomes the crux of one of Derrida's riffs on the way in which we both do, and do not, share a common world with other living beings, including and perhaps especially other human beings.

Derrida suggests that although we may not know what it means to inhabit or to cohabit, we do know that all living creatures die and therefore are mortal. Furthermore, although all living things may not inhabit or cohabit the same world, they are all finite. In spite of the differences between worlds—so many unbridgeable islands—Derrida asserts that there is one thing about which we can be certain, or at least believe, namely, that all living beings die.

Life is defined in terms of death, a definition that Derrida comes to associate with Heidegger. Derrida again takes up the question of what animals and humans share and answers "that all living beings, humans and animals, have a certain experience of what we call death" (11). We may not share a world, but all living beings share mortality. And, although we may not know how to define death, "we can believe that these living beings have in common the finitude of their life, and therefore, among other features of finitude, their mortality in the place they inhabit, whether one calls that place world or earth" (10).

Although in the many iterations of the line from Celan that mark Derrida's seminar he often mentions birth, it is death that comes to inhabit it. Thus it is through the Celan fragment that death inserts itself into the world. Through the Celan fragment, we have moved from Heidegger's discussion of the relation between world, finitude, and solitude to a discussion of death, dare we say, as such. It is interesting, however, that Heidegger analyzes finitude in terms of solitude rather than in terms of death per se. For Heidegger, the fact that we are finite makes each one of us unique. And, as Derrida points out, Heidegger knowingly equivocates on the meaning of solitude as being alone, which can mean either without company or without equal as in exceptional. Dasein is alone in terms of being the only being with access to the world (or Being) as such. Indeed, what sets apart Dasein as world-building is this uniqueness or solitude, what Heidegger sometimes calls individuality (e.g., Heidegger 1995 [1929–1930], §39). For Heidegger, stones are worldless because they are not unique individuals—all stones are alike. And while animals are more individuated than stones, still they are not unique individuals in the way that human beings are, at least according to Heidegger. In *The Fundamental Concepts*, however, it is not death—or a relationship to it—that determines whether or not one has a relation to world as such. In fact, it is uncanny how little Heidegger speaks of death in a text that includes finitude in its title. Yet perhaps it is more uncanny that the *world as such* is taken over by *death as such* in Derrida's retracing of Heidegger's steps, particularly insofar as Derrida criticizes Heidegger for choosing this very path (cf. Derrida 2011, 90–91). For, Derrida's next mention of the Celan fragment in his seminar appears as an indirect translation of the Novalis fragment. Derrida identifies the Novalis line—which he paraphrases as "philosophizing as an experience of nostalgia, as philosophy suffering from a constitutional sickness that would be homesickness"—

with Robinson Crusoe's nostalgia (at first for England and eventually for his island) "for the world he has lost (die Welt ist fort, as Celan would say)" (32, second parenthetical remark in original). We could say, translated into Celan's language ("as Celan would say"), Novalis's and/or Heidegger's, nostalgia becomes not only the seemingly unbridgeable distance from the world ("the world is far away") but also an ethical obligation to the other ("I must carry you"). Indeed, both the turn toward death and the turn toward ethics in *The Beast and the Sovereign, Volume II* revolve around this one line, the last line, from Celan's poem "Grosse, Glühende, Wölbung" ("Vast, glowing vault").[13] This line of poetry, repeated in nearly every session, is not only the axis around which Derrida binds the unlikely duo Crusoe-Heidegger but also it is a performance of a certain poetic world-making that Derrida proposes as a counterbalance to sovereign world-building. Poetic world-making echoes the distinction Derrida drew in *The Beast and the Sovereign, Volume I* between poetic majesty and sovereign majesty (cf. Oliver 2013) and, we might say, is resonant with the shift in Heidegger from world-building to poetic dwelling, analyzed in the last chapter. Derrida's turn from world to death marks both the necessity of the poetic and of the ethical.

DO ANIMALS DIE?

By inserting Celan between Robinson Crusoe and Heidegger, Derrida opens up the possibility of man's own worldlessness, if not akin to the stone's, just as absolute. Like and unlike the stone, "[w]e are *weltlos*" (2011, 9). Derrida focuses on the lack or privation of world in Dasein's unique capacity for world-building by turning his attention to Heidegger's profound boredom and melancholy. He finds an absent world in Dasein's melancholy solitude as the only being with this unique responsibility to carry the weight of the world. This privation is as profound as that of the animal's supposed poverty in world. Humans, then, are just as deprived of world as animals, if not necessarily in the same way. And, although Derrida is clear that "melancholy is not nostalgia," he says, "there is between these two affects an affinity, an analogy, that depends at least on the fact that these two sufferings suffer from a lack, a privation, even a bereavement" (111). Like and unlike the animals, we are deprived of world. Like and unlike stones, we are worldless. Ultimately, what renders us worldless and deprived is death, but not Heidegger's being toward our own death. Rather what renders us worldless is being toward the death of the other (even if, as we will see, that other happens to be one's self). Once we move from the self-centered being toward our own death to the other-centered being toward the death of the other, it becomes easier to "see" the effects of death and mourning in the animal kingdom.

If humans too are deprived of world, how can we be sure that we have access to the world as such and animals do not? How can we be sure that we have access to death as such and animals do not? For Derrida, these questions are intimately connected. Death and world come together. Derrida takes issue with Heidegger's confidence that we have access to the world, the Being and beings, and death as such, and that animals do not: "what seems more problematic still to my eyes is the *confidence* with which Heidegger attributes dying properly speaking to human Dasein, access or relation to death properly speaking and to dying as such" (116, my emphasis). As he does with other philosophers throughout the history of philosophy who are confident in dividing the world into humans and animals, Derrida challenges this certainty from both sides, the side of the animal and the side of the human. We could say that he exposes these philosophers as confidence men, con men, who have pulled the wool over our eyes about animals and our separation from them for too long now.

In *The Beast and the Sovereign, Volume II*, Derrida repeats and extends some of his analysis of Heidegger on animals from his earlier work. Here he focuses on the problematic notion that human beings have access to death *as such* while animals do not. Within the familiar double movement of both affirming that animals may have a relation to death and denying that humans have a relation to death as such, Derrida attacks the human-animal binary on many fronts (if not frontally, as he says). In addition to arguing that human beings are also worldless and deprived of world, Derrida suggests that on Heidegger's analysis we are driven by something outside ourselves, much like animals are supposedly driven by instincts or benumbed by their "disinhibiting rings" (e.g., Heidegger 1995 [1929–1930], 269–270).[14] Taking up the Novalis fragment "Philosophy is really homesickness, an urge [*Trieb*] to be at home everywhere," Derrida troubles this *Trieb* that drives us. *Trieb* can mean urge, drive or instinct; and on Derrida's reading, our very being is caught up—or *gripped* as Heidegger would say—by this drive. Derrida says that Dasein "is not only nostalgia, but a compulsive nostalgia, a drive (and Dasein is thus essentially a drive, a *Trieb*) that pushes it to be everywhere at home" (2011, 107, parenthesis in original). Dasein's *Trieb* or drive determines its relation to the world just as much as an animal's *Trieb* or instinct determines its relation to the world.

Derrida also challenges the circularity of Heidegger's argument that animals do not die. Of course, Heidegger admits that philosophy necessarily goes in circles. But, Derrida argues that this particular circle follows a "paradoxical sequence," namely, "having insisted on the fact that death, the moment of death, is the 'touchstone' (*Prüfstein*) of every question on the essence of life, here is Heidegger affirming that the animal cannot die, properly speaking, but only come to an end" (115). Clearly the animal is alive, and, according to Derrida, Heidegger defines life in terms of the ability to die; at the same time, he claims that the animal does not die, properly speaking. Heidegger insists that only humans are properly mortal (*Steirblich*) and, while human beings die (*Sterben*), animals only end (*Verenden*). It is at this point that Derrida calls into question Heidegger's confidence: "I hang on to this curious non-sequitur that consists in defining animality by life, life by the possibility of death, and yet, and yet, in denying dying properly speaking to the animal. But what seems more problematic still to my eyes is the confidence with which Heidegger attributes dying properly speaking to human Dasein, access or relation to death properly speaking and to dying as such" (116). To be fair, Heidegger, as always, gives different meanings to *death* and *die*. Moreover, what is at stake is to die, properly speaking, *as such*. In other words, for Heidegger, just as animals have a relation to the world but not to the world *as such*, they die, but they do not have a relation to *death as such*. Animals are mortal in the sense that all living beings are mortal, but, for Heidegger, only human beings realize that all living beings are mortal; only human beings have a relation to death as death.

At this point, Derrida moves his challenge to the other side, so to speak, now questioning how human beings have a relation to death as such. Just as in *The Animal That Therefore I Am* he is more interested in troubling this other side, the side of the human and our confidence in our own abilities, than in insisting that animals too have those abilities. Although he claims that, to various degrees, animals do have the numerous characteristics and abilities attributed to humans, he is more concerned to show that humans do not possess or own these very characteristics and abilities in the way that philosophers maintain that they do, for example, *as such*. He asks, how do we access death? It isn't enough just to say the word *death* or to see another's death or to imagine one's own death in order for one to access his or her death *as such*. Imaging one's own death is always an exercise in also imaging one's own survival. When we think of our own death, we necessarily imagine ourselves looking at our dead bodies. So we are split in two, both dead and surviving, viewing death from the vantage point of the living; we become other to ourselves. This leads Derrida to conclude, "our thoughts of death are always, structurally, thoughts of survival" (117). Furthermore, Derrida argues that at death—whatever that is—our bodies do not belong to us, but rather to the other or others. It is the other who decides what to do with my remains. My corpse/corpus is in the hands of the other alone (117). Derrida maintains that this lack of "habeas corpus," or having one's own body, begins before death; death is just the most extreme and radical case of the other having my remains, my body.

Through some poetic world-making of his own, *as if* following Robinson Crusoe and Heidegger, Derrida comes to settle on death as definitive of world. The loss of world is constitutive for "having" world. At the center of all world-building, then, is this loss of world, which in the end makes all world-building a fictional, if sometimes delusional, endeavor. Using a line from another poet, John Donne—who perhaps not coincidently is famous for saying "no man is an island"—Derrida suggests that the very structure of the Heideggerian *as such* (which distinguishes us from the animals) is one of loss and privation, even death. Derrida quotes Donne's "Holy Sonnets": "I run to Death and Death meets me as fast, / And all my Pleasures are like Yesterday." And he goes on to suggest various interpretations of this line, which culminate in the hypothesis, "pleasure is born only of the mourning, of enjoyment as mourning. And not any mourning and any memory of death, but the mourning of myself" (52). Again Derrida links the mourning and melancholy invoked in the poetic fragment to Heidegger's invocation of Novalis on philosophy as nostalgia or homesickness. He suggests that the structure of pleasure is that it is already in the past, already nostalgia for pleasure. Both death and pleasure are never present but only in some past future or future past, the future anterior, it *will have been* pleasure, it *will have been* my death. Without mourning and death, there is no pleasure: "Without mourning, and the mourning of myself, the mourning of my 'I am present,' there would be no pleasure. There would never even be an 'I am' a consciousness" (53). Pleasure as such, like death as such, cannot exist in the present, is never present, but always only the nostalgia for what is gone. It is gone before it is present.

The very structure of the *as such*, then, is nostalgic; it is always a look back at beings, from elsewhere, another time, another place away from home; the as such is homesickness itself. Put simply, in order to think pleasure—my pleasure as I experience it—I have to remove myself from it, and imagine it as such. Conversely, by so doing, I have killed pleasure. The moment that I start to think "this is pleasurable," pleasure is gone, and I am already mourning its passing. Insofar as my pleasure is from yesterday, as the poet says, then it is from the other and not my own: "My pleasure is from yesterday on, by yesterday altered, come from the other, the coming of the other" (53–54). The structure of desire is always from and for the other. Furthermore, returning to Robinson Crusoe, who cannot tell his footprint from that of another, we might ask: Can we ever be sure that our pleasure is our own and not that of another?[15] Even, as Derrida says, another myself—perhaps that one who took great pleasure in eating Captain Crunch as a ten year old. As Freud, Foucault, Derrida, and various feminists have taught us, in yet another sense our desires come from others, from the traditions and institutions of our culture. No man is an island.

"THERE IS NO WORLD, THERE ARE ONLY ISLANDS"

Yet, as we have seen, Derrida concludes his three theses, "there is no world, there are only islands," suggesting not only that every man is an island but also that every beast, and perhaps every living being, is an island, radically separated from every other. So how can we reconcile these two strands in Derrida's thought? Namely that the self is constituted through the other, on one hand, and that each singular being is unique to the point that its death is the end of the world, on the other hand. Again, Derrida makes a certain double movement in order to circumnavigate the seemingly oppositional, even contradictory, claims that no man is an island and every man is an island, that we are radically interdependent and that we are radically singular, that relationality permeates every living being and that alterity inhabits every living being. And again, rather than either/or, Derrida gives us both/and. Every living being is fundamentally relational and interdependent and yet, at the same time, each is singular and unique. Furthermore, the singularity of each comes through the other and others.

The answer to the question "do we share a common world?" then is both *yes* and *no*. We both share a common world and we don't. Each living being has its own world, and yet common codes (or languages, in the case of humans) assume that we share a world. Derrida argues that even a declaration of war assumes the possibility of peace insofar as it is a declaration addressed to another; all language, including declarations of war, comes from the other and others and is addressed to them. The Big-O Other is language or codes themselves, the realm of the symbolic and meaning that transcends every individual. The little-o other or others are those other beings who teach us codes and language and to whom we address ourselves, even when we are talking to ourselves. No matter what we say—including that there is no world or "the world is gone/far away"—we assume there is enough of a shared world that these statements will be heard and understood by another. In this sense, to echo Wittgenstein, there is no private language. As Derrida describes the tension between the constative and the performative dimensions of language in the case of declaring that we do not share a world, the constative indicates that there is no world while the performative assumes that there is one (2011, 259). Derrida argues that even the use of the word *world* in the phrase "the world is far away" assumes "that the addressee and the signatory of the statement share a language and comprehension of what 'world' means, inhabit the same world *enough* to be able to hear with one and the same ear and say with one and the same voice *Die Welt ist Fort*, so that the moment at which this phrase is spoken the world is still there" and that the two interlocutors "cohabit the same world" (259, my emphasis). It is noteworthy that even when insisting that we do share a world, at least *enough* of a world to assume that we can speak to each other, and perhaps understand each other, Derrida invokes images of war and loss of world, an idea to which we will return.

Traces, codes, rituals, tools, and tracks, in the case of animals, and languages, rituals, tools, and tracks, in the case of humans, operate as what Derrida calls "stabilizing apparatuses" (*dispositifs stablilisants*) or "prostheses" through which the fiction of the common or shared world is created. In his third thesis Derrida says, "the difference between one world and another will remain always unbridgeable, because the community of the world is always constructed, simulated by a set of stabilizing apparatuses, more or less stable, then, and never natural, language in the broadest sense, codes of traces being designed, among all living beings, to construct a unity of the world that is always deconstructible, nowhere and never given in nature" (8–9). In the very passage where Derrida sets out his most radical thesis, that there is no world, only islands, he also proposes that there is a common world constructed by stabilizing apparatuses such as codes and language, among others. No sooner has he said that the difference between one living being and another is unbridgeable than he says that we have constructed bridges, so to speak, that enable us to share a world, even if it is not "given in nature."

Just because something is not *given in nature*, however, does not make it unreal or nonexistent. Just because something is constructed, and therefore can be deconstructed, does not make it unreal or nonexistent. To the contrary, as we have learned from Derrida, perhaps above all others, the very appeal to *nature* and what is *given in nature* is a construct that can be deconstructed (cf. Oliver 2013). Furthermore, just because something is constructed and deconstructable does not mean that we cannot share it. In fact, just as any declaration is addressed to another, so is any construction, which means that these stabilizing structures assume a common world.

Of course, this is what Derrida is arguing, namely that our stabilizing structures or apparatuses assume a common world. In addition, he claims that they create the fiction of a shared world. Yet, we might ask, in the case of the world, isn't this fiction precisely what we mean by *reality*? If by *world* we mean the experience of living beings, certainly living beings share time and space with other living beings. Yet the world is constantly changing as new life forms are "born and die." Moreover, we have come to think that living beings themselves change and evolve through time and space. The interaction between living beings and their environment and each other is complicated, to the point that scientists constantly readjust their theories about the world and its reality. What is real? We can never be sure that we know. We can never be confident to the point of becoming con artists, conning ourselves into believing that our certainty is absolute. This kind of confidence easily becomes dogmatism and fundamentalism, which lead to a too often violent collision of worlds. Do we share the same world or the same reality? The answer necessarily will be *yes* and *no*. And saying that there is no world *in nature* merely displaces the question of whether or not we inhabit a common world onto the question of what appears in nature;

while saying that the world is a fiction merely displaces the question onto the question of what is the difference between fiction and reality.

Derrida's appeal to the assumption of a common world inherent in the performance of all signification, including codes andlanguages (and perhaps anything that can leave a trace), has more traction when it comes to thinking about whether or not there is a shared world. In other words, if deconstruction warns us against binary oppositions such as nature and culture, or reality and fiction, it also teaches us that the performative dimension of language can be at odds with the constative dimension in telling ways. In this case, any claim that the world does not exist—or that it is faraway or even that it is fictional—assumes that there is a world we share, one in which such statements make sense. Whatever the status of the claims themselves, their utterance presupposes that we do share the world with others. Thus Derrida argues both that each individual life is singular and unique and that we share the world with other living beings.

AS THE WHEEL TURNS

As we have seen, Derrida goes further than this when he suggests that our experience always and only comes through the Other and others when he discusses the line from John Dunne "my pleasure is like yesterday." The reflective gesture that gives us our experience as such, insofar as that is possible, comes from the other, both in the sense of other people (little-o others) and symbolic systems of representation (Big-O Others). In terms of the world, Derrida argues that one's sense of oneself as a unique individual comes through what he calls the "prosthesis" of the world.[16] Discussing *Robinson Crusoe*, Derrida analyzes Crusoe's attempts to reinvent the wheel as a metaphor of sorts for every individual being making its way in a world that it did not choose, paradoxically making that world its own by following in the footsteps of others. The wheel becomes a figure for all autos, including automobile, auto-affection, automatic, autobiography, and even autoimmunity, which in turn becomes a figure for deconstruction itself. The wheel, says Derrida, "turns on its own [*toute seule*] . . . by turning on itself" (2011, 78). This phrase has many meanings that are appropriate to Derrida's analysis of the ipseity of each individual that comes through its relation with others. First, to say that it turns on its own could mean that it moves itself as in auto-motion, automation, or automobile. It also means that it turns by itself, all alone, or that it turns itself on, as in auto-affection. But it does so by turning on itself, which could mean again that it turns itself on. Or it could mean that it turns on itself in the sense of turning against itself such that every auto-affection is also an auto-destruction. In other words, its automation is also an auto-destruct sequence. As the wheel turns, the ipseity of self is put into motion as self-moving, auto-affecting, and ultimately auto-destructing.

The wheel, then, comes to stand for this revolution from automation to auto-destruction that Derrida identifies with autoimmunity. With autoimmunity the immune system that is supposed to protect the organism turns on itself and destroys it. In a sense the organism becomes allergic to itself. Derrida sees this autoimmune self-destructive logic at work in both *Robinson Crusoe* and Heidegger's seminar, particularly in relation to the divide they set up between animals and themselves as human, the beast and the sovereign. It is as if there is a mechanism in the text that turns on itself. That is to say, the same mechanism that turns on the text turns on it, or the axis around which it revolves is also that which destroys it. In this case, that axis is sovereignty. The return to self that is the axis around which these texts revolve involves a loss of self, even a destruction of a certain sovereign self.

For example, Robinson Crusoe attempts to make himself independent from his parents and homeland by going to sea. Through his voyage he asserts his individual sovereignty. When shipwrecked, he attempts to reinvent himself to suit his new surrounding, but he does so only by holding onto the very traditions that he sought to escape, evidenced by his repeated invocations of his parents' warning and admonitions and his turn to the bible and prayer. Derrida reads *Robinson Crusoe* as a lesson in learning how to pray or, more precisely, learning yet again how to pray.[17] Derrida claims that the very paternal law Robinson Crusoe is trying to escape returns to him both to protect and to threaten him. The law of the father returns as protection insofar as Robinson Crusoe attempts to recreate it on the island with himself as the sovereign (insisting on wearing clothes and detesting cannibalism so as to maintain his status as a civilized Englishman); it returns to threaten him in the figures of the earth, the wild animals, and the savage cannibals that threaten to devour him alive (85). Just as Robinson Crusoe's invention of the wheel (a potter's wheel) is a reinvention, so too his invention of himself is a reinvention in the sense that his sense of himself and his actions always return to him from outside, from others, from the world (the "prosthesis" of the world). In this regard, the turning toward is always also a turning against one's self. Inventing and reinventing the self, then, at the same time, turns on undermining the self, insofar as that turning depends on the prosthesis of the world, which interrupts illusions of self-reliance and independence (85). Derrida goes on to discuss an uncanny passage in which Robinson Crusoe hears his parrot calling his name and at first doesn't recognize the voice as the psittacism of his own auto-appellation, which also foreshadows the uncanny moment when he is unsure of whether the footprint in the sand is someone else's or his own. Perhaps, like a wheel, he is just turning in circles. Robinson Crusoe comes to himself from the outside, from the prostheses that he creates (his tools, his talking parrot, his journal, his prayers and rituals).

In terms of Heidegger's seminar, Derrida suggests that there is the same autoimmune logic at work. Heidegger sets out to answer the question what is the world and, in order to do so, must lose the world. The reflective movement of the *as such*, taking the world as such, rips one from anything that might be immediate experience of, or in, the world, a move that leads to nostalgia and melancholy over what has been lost and cannot be regained. And yet Derrida argues that every return to the self necessarily follows the destructive logic whereby finding the self requires losing the self. The supposedly sovereign self necessarily erects it sovereignty by dividing itself and thereby undermining its sovereign unity. On the one hand, the sovereignty of the self is created only through interaction with the world and is therefore not sovereign but in fact dependent upon the world. On the other hand, the evidence of sovereignty appears only in the world as the products or remains of the self and therefore the self is again dependent upon what is outside itself. In other words, the very operations that construct the sovereignty of the self, its ipseity, also destroy it (see 2011, 88). Individuality, ipseity, comes through the world and the prosthetics through which it constructs that world. The autonomous, automotive sovereign self is split in its origin (and therefore neither so autonomous nor automotive as it thinks). If the world is the product of stabilizing apparatuses, so are individuality, ego, and even cogito.

While it seems clear from Derrida's analysis that our sense of ourselves as autonomous individuals comes through these stabilizing structures and apparatuses, what is less clear is the relationship between the production of the illusion of sovereign autonomous individuality and the singularity of each unique living being. Significantly, the former is antithetical to ethics while the later is the beginning of ethics. If the prostheses of and in the world shore up the ego, doesn't the singularity of each put it beyond these shores and separate it such that "there is no world, there are only islands"? How can we reconcile the tension between Derrida's insistence that each living being is singular and unique—an island—and at the same time that ipseity always comes through the other or the world? Indeed, what does Derrida mean by "world" if he can say both that there is no world and, at the same time, claim that each individual returns to itself through the world? Or, as he says, that the performative utterance "there is no world" assumes a common world?

Michael Naas answers this question by distinguishing between singularity as a unique opening onto the world and individuality as the ego or ipseity that closes off the world (Naas 2014). While each singularity is radically separated from all others, individual identity is a product of the stabilizing apparatuses or the prostheses of the world. We could go further and say that the ego or individual identity is itself a stabilizing apparatus produced like other prostheses in order to artificially support it. Naas suggests that the difference between singularity and individuality/ipseity is the difference between opening onto the world from the point of one's singularity and closing oneself off to the world from within the illusion of self-contained autonomy.[18] When Derrida invokes figures of islands, he refers to singularity as it opens onto the world, whereas when he analyzes ipseity he refers to the individuated ego as it closes itself off to the world.

For Derrida, what we might call our everyday notion of ourselves as individuals with constant identities over time is produced through the prostheses of the world; this identity returns to us through the world and its stabilizing structures, especially through language (cf. Naas 2014). Thus insofar as each individual identity comes through the common world it is neither unique nor singular. These identity structures are the same for everyone. But what is irreplaceable and singular in each living being is "ultimately a certain relation to time, to an unrepeatable, unique time and thus a relation to an unforeseeable future" (Naas 2014, 51). We might add that this singular relation is not only to time but also to place or location such that each living being "occupies" or "has" a unique experience of the world. Although each of us relies on the stabilizing structures of language, tradition, and rituals to think of ourselves as self-identical, at the same time, each of us has a unique position in relation to those structures insofar as each of us occupies a different time and place. No two beings are in the same place at the same time. So it is not that each being is unique in its substance, but rather in its location in time and space. Resonant with an Arendtian emphasis on natality, Derrida says, "What is absolutely new is not this, rather than that; it is the fact that it arrives only once. It is what is marked by a date (a unique moment and place), and it is always a birth or death that a date dates" (Derrida 2002, 104).[19]

Analyzing these passages from Derrida in which he speaks to the uniqueness of the time and place of each living being insofar as each is born (hatched, spawned, etc.) and dies, Naas argues Derrida's claim that there is no world, only islands is not solipsistic when we realize the distinction between individual identity and unique singularity as the difference between the "I," or ego, and what we might loosely call lived experience, or witnessing to that experience. More precisely, we have no access to the experience of the other except through testimony to it by the other who says to us, "believe me," this is how I feel. Since we cannot perceive in a direct way how this other feels—what goes on inside her head—we must take her word on faith. Appealing to Derrida's use of "faith" and "miracle" to describe this interruption from the other that constitutes the social bond, Naas argues, "everything in the world can perhaps be accounted for by the laws of nature or causality except this appeal ["believe me"], which opens up the world like a miracle and can only be affirmed through a kind of elementary faith" (Naas 2014, 52). And this faith in others is the glue that holds together the social bond. The "miracle" is that we can build bridges between islands through testimony and communication, which is always a response to the interruption of/from the other. It might seem that relying on testimony as the interruption from the other that connects one singular being to another excludes nonhuman animals and other life-forms. But Naas suggests that testimony can mean any trace of another, including traces left by nonhuman animals.[20] Every living being leaves a trace, a kind of testimony, which is addressed to and can be received by another.

THE END OF THE WORLD

But, what does it mean that each birth is the origin of *the* world and each death is the end of *the* world? What does Derrida mean that each life is not just *a* world but also *the* world? In his forward to the collection of eulogies for various friends, entitled *Chaque fois unique, la fin du monde* (Each/every time unique, the end of the world), Derrida announces at the beginning: "Each/every time, death declares the end of the world in totality, the end of each/every possible world, and each/every time, the end of the world as a unique totality: thus irreplaceable, and thus infinite" (Derrida 2003b, 9).[21] And he concludes, "death itself, if such a thing exists, leaves no place at all for the slightest chance, nor for a replacement, nor for the survival of this one and only, unique world" (2003b, 11).[22] Again, he ends with the Celan fragment, *Die Welt ist fort, ich muss dich tragen,* indicating his own obligation to carry his lost friends (or perhaps their remains) after the end of the world that is the end of each of their unique lives.

Throughout *The Beast and the Sovereign, Volume II,* Derrida repeats this idea that each death is the end of *the* world, the whole world, and not just the end of *a* world. One of the most striking passages comes late in the seminar, after the U.S.-lead invasion of Iraq in 2003, when Derrida insists that all war is world war and each lost life is the end of the world:

> And in every war, at stake from now on is an end of the world . . . and at stake is an end of *the* world (*Die Welt ist fort*), in the sense that what is threatened is not only this always infinite death of each and every one (for example of a given soldier or a given singular civilian), that individual death I've often said was each time *the* end of the world, *the* end, the whole end of *the* world (*Die Welt ist fort*), not a particular end of this or that world, of the world of so and so, of this one or that one, male or female, of this solider, this civilian, this man, this woman, this child, but the end of the world in general, the absolute end of the world—at stake is the end of the world (*Die Welt ist fort*) in the sense that what is threatened, in this or that war, is therefore the end of the world, the destruction of the world, of any possible world.
> (2011, 259–260)

There are several rhetorical elements that are significant in this long passage. First, there is Derrida's insistence that each death is the end of the world and that war threatens the destruction not just of a world among others, but of the world and any possible world; in other words, war always threatens to end lives and thus worlds, even *the* world, but now, with the threat of global war, war threatens to end the whole world and any possibility of world. The rhetorical force of the passage, however, comes from the repetition of the Celan poem fragment, which appears as a refrain, repeated three times. In the paragraph following this passage, he quotes it a fourth time, "more than ever *Die Welt ist fort*, [that] the poets, more than ever, more rare than ever, are more touched by truth than the politicians, priests and soldiers" (260). As we continue to see, Derrida's analysis of the world turns around this line from the poet Celan. Here again he links this fragment to our ethical obligation in the face of a disappearing world: "what there is to bear, as the responsibility of the other, for the other, must be borne where the world itself is going away" (260). Shortly, we will return to the ethical obligation suggested by poetry. But, for now, let's explore how the connection between the world and death has become a connection between the world and war, as if only by imaging the end of the whole world can we think of the world as a whole or as such.

Like the shots of empty streets and deserted cities that represent the death of all of life on earth at the end of Stanley Kramer's 1959 nuclear apocalypse film *On the Beach*, we can only imagine the end of the world, whether of our own personal world or the whole world, as postapocalyptic survivors watching what happens after the end. Just as the camera filming after the end betrays the existence of a survivor, our imaginary images of our own death or the end of the world on display in a film make manifest the structural necessity of the position of witness or spectator in all our thoughts of our own finitude. In Freudian terms we might say that our own death becomes a fetish of sorts insofar as we both believe in it and do not believe in it at the same time. Every invocation of our own death takes on the "as if" structure of the fetish: I know that my death is inevitable, but I act as if it is not. For Derrida, what he calls "the phantasm" of this "living death," associated with Robinson Crusoe's terror of being buried alive or swallowed up by beasts or cannibals, is "the inconceivable, the contradictory, the unthinkable, the impossible," namely the end of the world, which is, at the same time, "named, desired, apprehended" (148). This "as if" can be either a denial through which we arrogantly give ourselves the "right to the world as such," or, as we will see, it can be a poetic and creative gift of the world for another (cf. 260, 268).[23]

Derrida argues that what is at stake in all contemporary wars is "the institution and the appropriation of the world, the world order, no less"

(259). As we see in this long passage, he maintains that "the interpretation and future of the totality of beings, of the world and the living beings that inhabit it" are at stake in every war. All war, then, is ideological, in that the combatants are not just vying for territory or a certain part of the earth but also for their way of life and beliefs, which is to say their world. In Arendtian terms we could say that all contemporary wars are "total war" insofar as they are battles over whose worldview will dominate. Holy wars spurred on by fundamentalist doctrines and slogans are battles over how to interpret the world and which world order will prevail. This is to say that what is at stake in war is the answer to the question "What is the world?" All wars, then, are world wars.

Parties to these wars speak and act *as if* they can answer that question once and for all, *as if* they have a right to the world as such. They act *as if* they have the right, as Arendt says of Eichmann, to determine with whom to share the world and the earth. Indeed, the very notion of the *world as such* can play into the rhetoric of one homogeneous worldview or world order for the totality of beings as such. This is not, however, what Heidegger means when he talks about the world as such, which is a structural possibility for Dasein. Still, this structural possibility to take the totality of beings as the whole of the world can easily become the *as if* of denial that disavows the vast array of differences that might fill in the content of that empty structure. The *as if* of denial works along with the illusion of sovereignty as mastery that Derrida has worked to undercut throughout the seminar: we act *as if* we were the masters of our own destinies and the lords of the totality of beings; we act *as if* we can control ourselves and other beings, especially nonhuman being, but including other human beings, through violence in the name of our own worldview masquerading as the whole world as such. We act *as if* we can control our own violence and put it in the service of lofty ideals in the name of which we destroy worlds in order to save the world from them. In other words, threatened by worlds and worldviews that we do not share, we annihilate them in the name of our own world, which we insist is THE world, the one and only world. We deny both the multiplicity of worlds, what Arendt calls plurality, and the singularity of each in the name of "universal" principles that become the rallying cries for war, for example, "democracy," "freedom," and even, possibly, "cosmopolitanism."

When man believes he can master the forces of the universe, or master his own violence, he becomes a stranger to himself, a stranger to his own

Unheimlichkeit. This is when the hypothetical (fictional) *as if* becomes the dogmatic (metaphysical) *is*. And this is the ultimate weapon of mass destruction (insofar as it is behind the deployment of most, if not all, others). If, as Derrida suggests, there is no world as such, then how might we creatively pursue the world in ways that avoid the dangers of fundamentalism? In Derridean terms we might ask, what is in excess of sovereignty? Given that sovereignty itself is excess—the most, the highest, the absolute—what does it mean to think of something other in excess of excess (cf. 279)? Is there any form of sovereignty that could work against the violence of both state sovereignty as the right to let live or put to death and individual sovereignty as mastery? Relatedly, in terms of the problematic of the world, we might ask, "if there is no common world, but only islands, how can we communicate or build bridges between islands"? This question is especially relevant if the assumption of one universal world is part and parcel of dogmatic fundamentalisms. How, then, are we to navigate this quandary wherein we both need a shared world for the sake of communication and living together, cohabiting, and we need to acknowledge that each living being is singularly unique such that each one inhabits its own world or, more radically, is the world? Again, Derrida insists that rather than either/or—either we share a common world or we don't—the answer is both/and—we both share a common world and we don't. Each singular being is an island radically cut off from every other and yet we build bridges all the time, particularly through language, broadly speaking (so broadly that it can include the codes, rituals, and tracks of nonhuman life forms).[24]

Language, broadly construed, both assumes and provides a common shared world, a world created as a protection against the anxiety that comes from the realization that there is no common world and no common meaning. Paradoxically, then, language both makes and unmakes the world. It makes the world insofar as it provides the stabilizing apparatuses through which we bridge islands and create the illusion of a common world. It unmakes the world insofar as its untranslatability and dissemination undermine its stabilizing function and open up the world to an infinite variety of possible worlds. Derrida calls pretending or acting as if we give the same meaning to signs or words, including the word *world,* a "life insurance policy," "an agreement inherited over millennia between living beings," "an always labile, arbitrary, conventional and artificial, historical, non-natural contract, to ensure for oneself the best, and therefore the longest *survival* by a system of life insurances counting with probabilities and including a clause that one *pretend*, that one make *as if*, signing the insurance policy" (267). We sign on to the belief that we all inhabit the same world, the one and only world, as a life insurance policy. As it turns out, however, perhaps like all life insurance policies, it operates according to an autoimmune logic whereby what is supposed to protect us also destroys us. As we have seen, the belief in one and only one world, or even a common world that allows us to live together and cohabit, also can lead to war when an individual or a state gives itself the right to the world as such, which is to say, the right to determine the future and meaning of the world.

MAKING AND UNMAKING THE WORLD, POETICALLY

Again, we must ask, is there any alternative to sovereign world-making that becomes world taking? Derrida indicates that there is an alternative form of sovereignty that is in excess of political sovereignty with its superlative power. Poetic sovereignty, what in *The Beast and the Sovereign, Volume I*, Derrida calls "poetic majesty," works against political sovereignty and the will to mastery.[25] As he points out, *majesty* is from the Latin *majestas*, which means sovereignty (2011, 82). Thus Derrida uses one form of sovereignty against another, poetic majesty against sovereign majesty. Whereas sovereign majesty erects itself as the most, the grandest, and the supreme power, poetic majesty opens onto an uncanny otherness that unseats any such self-certainty. Poetic majesty or the majesty of art is used against political majesty to show how political majesty is itself an art form, a performance, or a fiction. Poetic majesty opens itself up to the *as if* in the absence of any absolute and certain world. Unlike political sovereignty and the will to mastery, poetic sovereignty avows rather than disavows the *as if*, which is to say, the fictional status of the world it creates.

In *The Beast and the Sovereign, Volume I*, Derrida goes further and suggests that poetic majesty gives us a chance, however precarious, however slight, of avoiding the deadly self aggrandizing fiction of political sovereignty that presents itself as The Truth of the World.[26] Poetic revolution disrupts the time of political sovereignty by giving time to the other, the time of the uncanny, which unsettles self-certainty of any "I can." Whereas the performance of political sovereignty claims to possess the power of the "I can" master the world, the performance of poetry undoes the sovereign "I can" through the ambiguity and necessary openness of language and interpretation that make multiple worlds possible. The world itself becomes the product of poetic majesty: the world as a poem calling out for interpretation and reinterpretation. This poetic revolution in the time of the living present ruptures the present as self-presence and reveals an absence at its heart, the absence of the world as such. For example, science can be seen as a search for the truth, or it can be seen as an ongoing interpretation of codes—what we might call the poetry of nature.

Reminiscent of Julia Kristeva's revolution in poetic language, Derrida's poetic majesty reveals the performative dimension of language operating even in proclamations of sovereign power or declarations of war. For Kristeva, the revolution in poetic language happens when poetic forms of language display their means of production, including their performative dimension and their materiality; that is to say, poetic language puts on display the very structures, forms, and techniques it uses to create a world. As we have seen, Derrida argues that there is a performative "pearl" at the heart of the constative oyster in any declaration of war that assumes the possibility of peace through a common or shared world and language (2011, 259). Language itself becomes the possibility of the world, shared or not; language itself becomes the bridge between worlds, a precarious and shifting bridge, never secure, but possible. Language, broadly speaking, "lightly" crosses the uncrossable difference and provides a bridge between one island and another, however light and fragile this bridge may be (cf. 267). Perhaps we should say addressability and response-ability rather than language since usually we think that language is reserved for man alone. Addressability and response-ability are not unique to man, but are manifest in various ways in most, if not all, living beings. Most, if not all, living beings respond to others, and many also address themselves to others. Elsewhere I call this address and response structure *witnessing*. More recently, Cynthia Willett has namedit "call and response" precisely in order to conjure nonhuman voices (Willett 2014, 9–17).

Calling Celan as his witness, Derrida claims that *the world* is a word, a convention, that does not exist as such, and when those conventions disappear the world is gone and I must carry you (cf. 2011, 266). In nearly every, if not every, place where Derrida talks about the end of the world, or the death of each singular living being as not just the end of *a* world but of *the* world, he invokes this one line from Celan. Indeed, he appeals to this bit of poetry to make his case so many times that in most places he need only use a fragment of the fragment, one clause or the other, or just one word, *fort* (*gone*) or *tragen* (*carry*), to conjure the entire fragment with its world so far away that I must carry you. In fact, he cites all or part of the Celan fragment no less than forty-four times throughout the seminar, in nearly every session. How does this one line, from a poem that Derrida never cites in its entirety, come to stand in for the world and its absence?

To answer that question, we need take a detour through Derrida's analysis of the entire poem in "Rams Uninterrupted Dialogue—Between Two Infinites, the Poem," his homage to Hans-Georg Gadamer (2005b). There Derrida invokes this same line at the beginning of the second section, after discussing his "dialogue" with Gadamer and ending the first section with melancholy musings on the death of a friend as the nature of friendship wherein one friend is doomed from the start to face the death of the other. Derrida says that the survivor is left to "carry the world of the other, which I say without the facility of a hyperbole. The world after the end of the world" (2005b, 140). Given Derrida's sometime appeal to what he calls "hyperbolic ethics," it is significant that he insists that calling the death of each unique living being "the end of *the* world" and not just the end of *a* world is not hyperbole.[27] This is not a gesture of hyperbolic ethics. Without hyperbole then, Derrida contends, "each time, and each time singularly, each time irreplaceably, each time infinitely, death is nothing less than an end of *the* world. . . . Death marks each time, each time in defiance of arithmetic, the absolute end of the one and only world, of that which each opens as a one and only world, the end of the unique world, the end of the totality of what is or can be presented as the origin of the world for any unique living being, be it human or not" (140).

In this passage, Derrida claims that each death is unique, that it is the end of *both a* world ("not only one end among others") and *the* world and not merely something or someone in the world. He maintains that each death marks in time *the* end of the world, not just the end of *a* world or a life. Recall that elsewhere Derrida describes the singularity of each life as marked by time, more specifically, by a date or two dates: a birth date and a death date. These dates and times mark this life as unique. The phrase in the passage just quoted, "Each time in defiance of arithmetic" suggests not only that the time of life and death defies the clock time of dates and times on birth and death certificates but also that deaths are immeasurable, incomparable, and thus not a matter of death tolls.[28] One is enough to make—or take—the whole world. Returning to our question to Arendt, how many does it take to make up a world or plurality, Derrida would answer, only one. Each living being offers a unique opening onto the one and only world. And yet each death also brings the end of "the totality of what is or can be presented as the origin of the world" (140). One way to interpret this claim is that each death interrupts the world and renders inoperative its stabilizing structures, that is to say, what makes it both *a* and *the* world. Faced with the death of a friend or family member, the structures that had made sense of the world, that had allowed us to live in it together, disappear. In this sense we face each death alone. Derrida says, "The survivor, then, remains alone. Beyond the world of the other, he is also in some fashion beyond or before the world itself. In the world outside the world and deprived of the world" (140). Rather than guarantee our access to world, death, death as such, shatters the world and leaves us, perhaps like Heidegger's animals, or even his stones, deprived of world.

This is the point at which Derrida makes the ethical move and again calls Celan as his witness. When the stabilizing apparatuses that hold the world together break down and death renders them inoperative, there are no words, rules, morals, rituals, or traditions that can support the weight of death. The survivor must fend for himself. And yet, in this worldless place, the nonplace of facing the death of the other, the survivor must carry that weight himself. He is responsible for carrying the other forward in this worldless world. He is "assigned to carry both the other and *his* world, the other and *the* world that have disappeared, responsible without world (*weltlos*), without the ground of any world, thenceforth, in a world without world, *as if without earth* beyond the end of the world" (140, my emphasis). Without ground, *as if* without earth, to stand on, responsibility carries on. Without world, *as if* without earth too.

Thus, if worlds are associated with cultural traditions, rituals, and conventions (whether human or nonhuman), earth is associated with our embodied connection to life and home. In other words, to be without world is to be without the codes and mores that govern our societies. To be without earth is to be without the very conditions of possibility for life itself. If ethics begins where the world ends, this means ethics is beyond the conventions of culture. But to say that we are *as if* without earth is to say that we are *as if* without the very bodies and what sustains them that keep us alive. To be without world is to step into the void where ethical decisions cannot be made based on accepted rules or conventions. If decisions are made merely in terms of "following orders," or acting on laws or traditions, then they are unthinking. When taken to the extreme, we end up with Eichmann. Ethics, then, is always necessarily beyond world; at least it requires acting *as if* we had no world to fall back on, imaging that the world is gone and I must carry you. But even ethics cannot take us beyond earth. For what would it mean to act *as if* we did not have bodies and were not earthlings? More to the point, what could be the meaning of ethics in this science fiction fantasy? What does ethos mean disconnected from earth.

Thus, the *as if* in "as if without earth" signals a turn to hyperbole, even to hyperbolic ethics. If world is associated with human worldviews and human conventions, we can imagine a world without world. But, given that all life exists by virtue of the earth, we can only imagine *as if* without earth. In order to conjure an image that speaks to this terrible responsibility, even impossible responsibility, we imagine ourselves not only worldless but also *as if* without the earth itself, which signals the hyperbolic ethical turn. This poetic *as if* makes all the more vivid what it means to face the death of the other without anything to support the weight of the world that bears down on the survivor who must bear it. Enter poetry, specifically Celan's "Die Welt ist fort, ich muß dich tragen" (140). Immediately after invoking the *as if*— "as if without earth beyond the end of the world"—Derrida again appeals to Celan, *as if* the ethical turn needs to call the poet as witness: more precisely, the poet who witnessed the end of the world that was the Holocaust.

The poem "wanders," says Derrida—perhaps like a planet, since as he reminds us, the Greek *planetes* means wandering—"but in a secretly regulated fashion, from one referent to another—destined to outlive, in an 'infinite process', the decipherments of any reader to come" (146). The poem is like any trace in that it is cut off from any original meaning or authorial intent, "an unfortunate orphan," which "always remains an appeal (*Anspruch*) to the other, even if only to the inaccessible other in oneself. . . . Even where the poem names unreadability, its own unreadability, it also declare the unreadability of the world" (147). Derrida suggests that through this wandering from one meaning to another within the universe of language (the Other) and moving among possible future readers (the others), the world opens onto this radical undecidability that they both share—the poem and the world. And this is the gift of thinking, the gift of the poetic *as if*, as if the world has constantly renewable meaning. Neither the poet nor the poem is sovereign over the possible meanings of the poem. Rather, it is destined to wander, not aimlessly or without meaning, but without any set telos or fixed meaning.

In this regard, the poetic *as if* is not the *as if* of a Kantian regulative ideal (cf. Derrida 2011, 269).[29] It is not aiming for perfection or the one and only true meaning or correct interpretation. It is not perfectible in the Kantian sense of getting better. So, too, the poetic *as if* is neither the *as if* nor *as such* of philosophy. It is neither the analogical *as if* nor the metaphysical *as such* of Platonic or Heideggerian philosophy.[30] Derrida positions the poetic *as if* as a counterweight to the philosophical *as if* insofar as the poetic as if avows its fictional status rather than disavows it. Unlike the philosophical *as if* that presents itself as the truth—the one and only truth of the world—the poetic *as if* puts on display the allegorical, mythical, and fictional process of making worlds—even the world—through signifying systems, what Derrida called *stabilizing apparatuses*. Unlike the philosophical *as if* and *as such* that claim to lift the veil and reveal the world as it is, the poetic *as if* suggests a wandering truth only through veiling; the shroud is the world, and, if anything, it shows that what lies behind or beyond is absence, lack, and void. Like the philosophical *as if* and *as such*, poetry steps into this void, but, unlike philosophy, rather than pretend to fill it, poetry supports the void and carries us through it.

The poetic *as if* allows us to act *as if* we inhabit the same world, *as if* cohabitation is possible. And yet poetry is also where this phantasm of cohabitation comes up against its limit insofar as it shows the multivocity of language and languages, the untranslatability from one island to another, which, at the same time, demands translation and interpretation. The poetic *as if* simultaneously displays the singularity of each and the absence of one common world and the possibility of bridging this abyss, however slightly and precariously, to create the possibility of a shared world or at least the possibility of address and response-ability that allows us to act *as if* the world is inhabitable and cohabitable. When we find ourselves worldless—as if without earth beyond the end of the world—the wandering of the poetic *as if* must do the heavy lifting in the face of this absent world. This does not mean that it *can*, but only that it *must*. Against the Kantian *ought* implies *can*, Derrida leaves us with an *ought* that not only does not imply can but also may imply cannot. Each one must attempt to translate what cannot be translated; in a sense, each one must attempt to translate the ethical into the moral, the unconditional into the conditioned, to translate the singularity and irreplaceability of every other, every living being, into universal principles (2005b, 162). Ethics must be translated into the world after the end of the world and where there is no world. The ethical obligation, then, is an impossible obligation, one that begins when the world ends. If it were possible, that is to say if it were doable and within the power of the sovereign *I can*, it would be moral rule following rather than ethical decision making.[31] In other words, it would be a thing of the world that can be understood and carried out by sovereign rationality and will, according to moral principles explicable to all rational beings. Yet, where there is ethical obligation, there is no world, but only the face-to-face relationship that obligates one singular being to another. This radical responsibility interrupts the world, it juts through the world and shatters the ground upon which we stand together. And in that moment my responsibility for you, for the other, is as singular as the other herself. Derrida describes this singular obligation without world or alibi: "As soon as I am obliged, from the instant when I am obligated to you, when I owe, when I owe it to you, owe it to myself to carry you, as soon as I speak to you and am responsible for you, for before you, there can no longer, essentially, be any world. No world can any longer support us, serve as mediation, as ground, as earth, as foundation, or as alibi . . . without earthly or worldly ground, the responsibility for which I must respond in front of you for you" (158).

To say that there is no world is to say that there are no moral codes, universal principles, common languages, rational structures, religious doctrines, traditions, or conventions to which one can appeal in the face of the ethical obligation to an other. The ethical obligation brings us face to face with the other, as Lévinas would say, and this encounter takes us beyond the world of stabilizing apparatuses. There is no apparatus, no machine, for making ethical decisions. Rather, the respondent is alone in his responsibility to the other and for the other outside of any possible world, without alibi. Living up to one's ethical responsibilities to the other means being alone and without world. For within the world we find morals and principles, but we do not find ethical obligation in the sense invoked by Lévinasian insomnia or Derridean hyperbolic ethics. Moral codes make it easier to do our duty by prescribing it for us. Ethical decisions, on the other hand, are those that keep us awake at night and force us to step out of comfortable rule following. Ethics demands that we vigilantly and continually try to avoid becoming merely answering machines and instead respond to the call of and from the other. Unlike morality that is decided once and for all—what is right is right for all time—ethics requires us to dwell in the undecidable space of the impossibility of knowing what is right and yet being obligated to do it nonetheless. Only then does responsibility become radical enough to open up the possibility not just of a world but also more importantly of a just world.

Ethically we are alone and yet not alone. Again, we have moved beyond the disjunction of either/or and moved toward the conjunction of both/and: we are alone and we are not alone. The hyperbolic urgency of ethics comes from behaving *as if* we are alone, without world, facing the decision of a lifetime and our own responsibility to and from the other. But, at the same time, we are by virtue of the other and others. There is no "I" without the ethical call from others or without the stabilizing apparatuses of the Other, most especially language. This is to say that the "I," the self, is a response to the call of others. And this response-ability comes from others and is then addressed to others. Still, Derrida's hyperbolic ethics of singular responsibility in front of the worldlessness of the world is not some heroic individualism whereby, like Hercules, we take on the weight of the world. And yet that weight is squarely on our shoulders.

"THE POET HOISTS NO FLAGS"

It is noteworthy that Derrida names his engagement with Celan, "Rams," after the animal named in the poem. As he points out, rams are often the sacrificial animals of choice throughout the Old Testament. More specifically, the ram's horn, the shofar, is associated with both the Jewish New Year and the day of atonement; thus Derrida proposes that the ram stands for both the beginning and the end of the world (2005b, 156). In Celan's poem there is an image branded between the ram's horns, perhaps the image of the poem itself, which is followed by a charging force, perhaps of the ram himself, which Derrida interprets as the rebellion of all scapegoats, all substitutes, all victims of holocaust, human and nonhuman (156–157). The charge of this sacrificial ram is aimed against its persecutors and against sacrifice itself, to the point, Derrida argues, of wanting "to put an end to their common world" (157). Perhaps like Hannah Arendt, who claims that no one wants to share a common world with Eichmann, this ram does not want to share the world if that world is one of holocaust.

Derrida interprets the poem's vast glowing vault of black stars as the home of the zodiac sign of the ram in the poem. Yet this vast glowing vault with the swarm of black stars pushing themselves out and away can also be read as an explosion, the big bang, perhaps the origin or the end of the world. The ram might be the destructive force resulting from this explosion that leads to the end of the world. The "I" may be the poem itself: "I brand this image, between the horns . . . I must carry you." "You" might be the poet himself, who relies on the poem to carry him after the end of the world, after the holocaust. The poem may offer protection, as slight as it is, against the destruction of the world. This is the way that Celan describes language. In his Bremen speech Celan suggests that language is what remains after every holocaust. It is what survives. Language survives the death of the other. And it is language that enables one to carry that other, perhaps carry him forward by interpreting his poetry. The poem too is dependent on the other (cf. 159). The meaning of the poem is entrusted to the other. Poetry, perhaps more than other forms of language, makes the need for interpretation apparent. It calls out for interpretation. And its meanings change with time and context. Poetry, like all language, is always addressed to the other, but its form makes the need for an active reader or listener more apparent than some other forms of language.

Celan describes his poetry as a movement toward the other, toward a responsive and responsible other that might make it possible to inhabit the world. If, as Celan describes it, language is what survives against loss, then it operates as a buoy, of sorts, on the seas of trauma and torment, eventually washing ashore, like a message in a bottle, in the hopes that the shore is inhabited and it will be taken up. Writing not only assumes habitation, it creates it; that is to say, it creates a world inhabited by others who can listen and hear. Thus Celan describes poetry as a going towards: toward this inhabited shore, toward the other.[32] In this sense poetry creates and recreates the witnessing structure of address and response that supports the possibility of subjective and intersubjective life.

Celan's poetry, then, can be read as an attempt to recreate an addressee in the face of the brutality of the Second World War.[33] Shoshana Felman goes further and argues that his poetry is engaged in the project not only of seeking a responsive listener but also of addressability itself. She claims that he "transforms poetry from an aesthetic art form into an inherent and unprecedented testimonial *project of address*" (1992, 38). Felman says that in Celan's poetry "the breakage of the verse enacts the breakage of the world" (25). In this broken world, as Derrida repeatedly insists, the fundamental relationship is one of responsibility: The world is gone, I must carry you. There is no sovereign "I can," but only an "I must." Where the world ends, responsibility begins. Resonant with Derrida's own analyses of Celan throughout *Sovereignties in Question: The Poetics of Paul Celan*, Hans-Georg Gadamer emphasizes the ways in which Celan's poems challenge the sovereignty of the subject. He describes the poetic word as a cosmic event that establishes its own authority beyond that of its author. The poetic word is one that even the poet cannot possess. For this reason, Gadamer says, "The poet hoists no flags" (1997, 221). Language has authority, but it does not belong to anyone. No one can claim the authority of the trace. If the trace "belongs" to anyone, it "belongs" to the other.

Immanuel Lévinas too sees in Celan a poet of otherness par excellence. Like Gadamer and Derrida, Lévinas suggests that Celan's poetry challenges the sovereignty of the self (1978, 19). If there is a return to self through the poem, it is as stranger, foreigner, or uncanny otherness. This strangeness is the strangeness of the other, of the self, and of man himself: "Strangeness of man—one touches man outside of all rootedness and domestication. Homelessness becomes the humanity of man—and not his denigration in the forgotteness of Being and the triumph of technique" (Lévinas 1978, 19). We could say that poetry helps us to be at home in our homelessness and, at the same time, makes what we take to be home and familiar uncanny. Positioning Celan's poetry as a counterweight to Heidegger's emphasis on the home of man in the house of Being, which is language, Lévinas embraces Celan as a fellow traveler on an "adventure where the I dedicates itself to the poem so as to meet the other in the non-place . . . a native land that has no need to be a birthplace" (1978, 20, my emphasis). Invoking Heidegger's notion of native ground, Lévinas suggests that this native ground need not become nationalism based on birthplace. Rather, poetry conjures a native land otherwise. Poetic language allows us to plant ourselves in language as a homeless home, a wandering that makes our home everywhere and nowhere. As a counterpoint to Heidegger's notion that man is the shepherd of Being, Lévinas finds in Celan, a Jewish Holocaust survivor, a "significance which signifies for the human as such, of which Judaism is an extreme possibility—to the point of the impossibility—a break with the naiveté of the shepherd, the herald, the messenger of Being" (1978, 20). For Lévinas, Celan's poetry signals a going toward the other that is a radical break with the isolated philosopher on a path through his familiar forest. This poetic wandering is not teleological; it does not have a goal or an end point. Following the metaphor of meridian in Celan's speech by that name, Lévinas describes the adventure of poetry as "the circularity of the meridian—perfected trajectory of this movement without return—, which is the 'finality without end' of the poetic movement. Poetic movement opens up the possibility of the infinite movement towards the other that creates the world *as if* I were reunited with myself there" (1978, 20).

For his part, in his few prose works, Celan talks of "the world-open uniqueness of great poetry" and a language of the earth that is "without I and without You," a language spoken by sticks and stones (1986, 57, 19–20). Two different languages: a language of the world and a language of the earth. The language of the world continually and repeatedly conjures *I* and *You*, while the language of the earth is beyond such distinctions, the language spoken by sticks and stones. And, "the language that counts here," is the language of glaciers, mountains and stones (Celan 1986, 19). Yet perhaps these two different languages are as essentially related as *I* and *You*, infinitely addressed to each other, forming and reforming the fragile map between islands, a map that leads us not where we want to go, but to the otherness of where we ought to go. Celan ends his speech, "The Meridian," describing what he finds following the map of poetry: "I find something as immaterial as language, yet earthly, terrestrial, in the shape of a circle which, via both poles, rejoins itself and on the way serenely crosses even the tropics: I find ... *a meridian*" (1986, 55). A meridian is an imaginary circle that bisects the poles of a celestial body from north to south and effectively slices it in two, like *I* and *You*, or earth and world. Like a poem, or language itself, the meridian is an imaginary marker we can use to orient ourselves on the planet, but not just any planet, on the earth. The earth is our native land, our here and everywhere, the only place that we can live. As Lévinas says of "Meridian," "native land on the meridian—which is to say: a here which is also the everywhere, a wandering and expatriation, to the point of depaganisation. Is the earth habitable otherwise?" (1978, 20).

Is the earth habitable otherwise? This question could suggest either that the only way to inhabit the earth is through poetry, or language as a wandering, or that we need to search for ways of inhabiting the earth otherwise,

otherwise than Being or otherwise than being the self-same and autonomous origin of sovereignty. What would it mean to inhabit the earth otherwise than sovereignly? Otherwise than lording over it as the masters of the totality of beings? Or even otherwise than the shepherds of being, tending it like a ram? Both our possible interpretations leave us homeless wanderers making our way on the earth by enveloping it in various worlds created through the fabric of language, not as a system of meaning so much as an address to the other, an address that conjures addressability itself. "I cannot see any basic difference between a handshake and poem," writes Celan to Hans Bender (1986, 26). The poem becomes a welcoming gesture that invites response. The poem as a form of hospitality that both assumes and creates the possibility for response-able cohabitation. The earth is only habitable with the addressability and response-ability from/to others with whom we cohabit, even those who don't (literally) read or write poetry. This habitability as cohabitability is not, however, unique to human beings. Animals too respond to others and the environment, and without their response-ability and the ability to address themselves, broadly speaking, to others, whether their fellows, other species, or their environment, the earth would not be habitable. In other words, the earth is habitable only by virtue of shared worlds, even if those worlds shift, change, and never completely coincide. Insofar as every living being leaves a trace, as Derrida might say, and insofar as many of them learn to read the traces of others, in a certain sense, they too dwell poetically. They too learn to interpret signs, make signs, and create worlds for themselves and others on the earth.

ETHICS BEGINS WHERE THE WORLD ENDS

It is telling that Derrida ends his analysis of Celan's poem in "Rams Uninterrupted Dialogue—Between Two Infinites, the Poem" with a turn to Heidegger's three theses on the worldlessness of the stone, the poverty of world of the animal, and the world-building of the human, commenting that "nothing appears to me more problematic than these theses" (2005b, 163). He goes on to ask what, if the world is gone, the *Fort-sein* of the world, exceeds any and all of these three possibilities. He suggests that the very thought of the world must be rethought "from this *fort*, and this *fort* itself from the '*ich muß dich tragen*'" (163). The world must be thought from the absence of the world as it relates to our ethical obligations to others, that is to say, our obligations face-to-face with others without the mediation of the world or any stabilizing apparatuses that tell us how to behave. We must act *as if* we are creating the world anew, reinventing the wheel. And, yet, to do so ethically demands that we do not create as lords or masters of the world or the earth, but rather, as caretakers who acknowledge our dependence on the other. How do we begin this task? Derrida answers at the very end of *Rams*, "I would have begun by recalling how much we need the other and how much we will still need him, need to carry him, to be carried by him, there where he speaks, in us before us . . . 'for no one bears this life alone'" (163).[34] No man is an island.

The end of the world is the beginning of ethics. This is because, unlike moral codes or laws, rules or doctrines, ethics is not in the world. Rather, it always exceeds the world. It is always a remainder left over after the world is gone. On Derrida's reading of the Celan fragment, ethics follows from the absence of the world. Indeed, in "Rams," Derrida reads the two clauses of the last line of the poem as a conditional phrase, *if* the world is gone, *then* I must carry you. But, he suggests that the conditional is reversible, if I must carry you, then the world is gone (158). In logical terms, this reversibility of the conditional if-then signals the most binding kind of relationship of necessary and sufficient causality. Yet insofar as these terms are not presented within a logical argument, but rather in an analysis of a poem, and more generally of the poetic as if, we might interpret the force of the bind as the ethical bind of a hyperbolic ethics. The urgency and vigilance required by hyperbolic ethics are signaled in this poetic *as if there were no world*, not even any earth. The groundless *I must* comes from here. Ethics begins where the world ends.

Following Lévinas, Derrida maintains that the ethical obligation comes before the ego or cogito. He says, "Before I *am*, I *carry*. Before *being me*, I *carry the other*. I carry *you* and must do so, I owe it to you" (162). On one level we could say that our sense of ourselves as individuals comes from the other. In terms of *The Beast and the Sovereign*, the ego and cogito come through the prosthesis of the world, through those stabilizing structures, most especially signification, through the Big-O and little-o others of meaning and other living beings. In a sense, we learn who we are and enter the world through these others. This is just as true for other creatures as it is for us. Each of us comes into the world anew, born for the first time, reinventing the wheel, so to speak.

Contra a certain Heidegger, Derrida suggests that our Being-in-the-World as world builders (*Weltbildend*) is not what is definitive of the uniqueness or solitary finitude of Dasein. Rather, it is our being worldless in the face of the other that brings us to the world as responsible for the world of the other in front of the absence of the world. This is not Heidegger's being-toward-death, which defines the finitude of authentic Dasein. Rather, it is being-toward-the-death-of-the-other, whether this other is the Other of the World and its stabilizing apparatuses or the death of another living being. Indeed, the death of another living being, one who is close, friend or family, whether human or beloved nonhuman, rents the world. Even the various traditions, religions, and rituals around death cannot translate or carry the weight of the loss, which Derrida insists is not *like* the end of the world but *is* the end of the world. With the loss of a loved one there is no world left to support such grief, yet there is still the responsibility to the other to carry him/her/it, perhaps only in memory.

This being toward the death of the other, however, is not unique to human beings alone. All living beings survive and are survived by others close to them. And, as we know, some nonhuman beings also have elaborate rituals around death and mourning; they have their own stabilizing apparatuses to cope with the end of the world that is the death of the other.[35] For example, it is now well known that elephants mourn their dead with elaborate rituals that go on for years: "When an elephant dies, its family members engage in intense mourning and burial rituals, conducting weeklong vigils over the body, carefully covering it with earth and brush, revisiting the bones for years afterward, caressing the bones with their trunks, often taking turns rubbing their trunks along the teeth of a skull's lower jaw, the way living elephants do in greeting" (Siebert 2006, 4; see also Nichols 2013). In addition, elephants "have a memory that far surpasses ours and spans a lifetime [seventy years]. They grieve deeply for lost loved ones, even shedding tears and suffering depression" (Sheldrick 2012). Last year, two herds of wild South African elephants traveled for twelve hours to visit the house of elephant conservationist Lawrence Anthony upon his death. Anthony's family has no idea how the elephants, most of whom he saved and then returned to the wild, knew he had died. But, within days of his death, two herds arrived within a day of each other, both in single file as if in a funeral procession, and stayed at the house, perhaps in vigil, for two days before returning to the bush.[36]

Elephants possess many of the characteristics traditionally reserved for humans, including memory, mirror self-recognition, self-awareness, emotions, attachment, social bonds, mourning, stress, trauma, communication, humor, and other complex behaviors and social structures.[37] Neuroscientists have established changes in elephants' brains after trauma, and they have discovered that elephants have an "extremely large and convoluted hippocampus that is responsible for mediating long-term social memory" (Bradshaw and Shore 2007, 432; Siebert 2011). Bradshaw reports that elephants display symptoms of posttraumatic stress that resemble those of humans, especially when as youngsters they witness the massacre of their families (Bradshaw 2004; see also Siebert 2011).

> Studies show that structures in the elephant brain are strikingly similar to those in humans. MRI scans of an elephant's brain suggest a large hippocampus, the component in the mammalian brain linked to memory and an important part of its limbic system, which is involved in processing emotions. The elephant brain has also been shown to possess an abundance of the specialized neurons known as spindle cells, which are thought to be associated with self-awareness, empathy, and social awareness in humans. Elephants have even passed the mirror test of self-recognition, something only humans, and some great apes and dolphins, had been known to do.
>
> (SIEBERT 2011, 54)

Some of these studies suggest that elephants have even better long-term memories, particularly social memories, than humans (cf. Sheldrick 2012). Given their capacity for memory and emotion, one researcher concludes, "more than any other quality, the elephant is thought of as having understanding of death. Grieving and mourning rituals are an integral part of elephant culture" (Bradshaw 2004, 147). Is it possible that elephants understand death better than we do? Perhaps it is they, and not us, who have a relation to death *as such*?

Derrida's analysis in *The Beast and the Sovereign* allows for this possibility even as it challenges the Heideggerian certainty that human beings have access to death as such. Unsettling any conclusions about the world as such, calling Celan's poetry as witness, Derrida repeats the litany of an absent world, far away, gone, that carries with its very absence the ethical obligation to each singular other living being as not only *a* world but also *the* world. Can we learn to share the earth with those with whom we do not even share a world? Perhaps by taking up our singular ethical responsibility to every living creature *as if* to the world itself—*as if* to the very earth itself—we may hope to learn to inhabit and cohabit the earth. Perhaps if we treat the death of each living being as the end of not just *a* world, but of *the* world, we may hope for peace, or at least the end of holocaust. If the absence of the world as such gives rise to ethical obligations to the other, we may hope to move beyond dogmatism and fundamentalism and towards what we might call poetic tolerance of different worlds. If poetic sovereignty, or the power of interpretation, replaces political sovereignty, or the power of the will, perhaps we might begin to give up our illusions of mastery and start to appreciate poetry in the codes, rituals, and tracks of each singular living being. For the power of interpretation is based on an acknowledgment that every interpretation is provisional because meaning is fluid and comes into being only through relationships, in the space and time between singular beings, and evokes the miracle of communication between islands or across worlds.

What happens when we return our gaze from the "vast, glowing vault with the swarm of black stars pushing themselves out and away" and back toward the Earth, which from that vantage point looks so singularly beautiful and yet so incredibly small? Even if "the world is gone," the Earth remains. Thus we can found the ethical import of "I must carry you" on the groundless ground of the earth, insofar as it remains our one and only home. In the whole of that vast, glowing vault, the Earth is the only planet that sustains us and every living being that we have heretofore encountered. If the singularity of each living being obligates us to respond to it in a way that opens up rather than closes off its world and its ability to respond, then certainly the singularity of our planet obligates us to respond to it in a way that opens rather than closes off response. Indeed, our singular bond to the earth and to other earthlings may be the groundless ground of ethics that not only moves beyond morality and politics but also nourishes them insofar as they offer us principles with which to share the earth as home and acknowledge that we belong to the earth, along with every other earthling. Perhaps the threat of disappearing worlds, whether human or nonhuman—a threat that is increasing exponentially because of man-made pollution—can return us to the Earth and an earthbound ethics by making us realize that even if we do not share a world, we do share a planet.

6

TERRAPHILIA

Earth Ethics

> The true frontier for humanity is life on earth
> —EDWARD O. WILSON, "BIOPHILIA AND THE CONSERVATION ETHIC"

In the haunting film *All Is Lost* (2013), the nameless protagonist (listed only as "our man" in the credits) is like an island, alone and adrift on the sea, the only person in the film—unless you count a school of fish and some sharks. This unlucky Robinson Crusoe does not find refuge when his boat sinks. Rather, the boat's sinking is just one of the harrowing accidents that befall "our man" (Robert Redford) as we watch his world slowly disappear piece by piece until there is nothing left but his body floating in the murky sea. While there are many ways to interpret C. J. Chandor's enigmatic film, it can be read as a parable of the dangers of man-made pollution and climate change and the manly, but therefore futile, attempt to control it. In this final chapter, brief discussions of three recent films, *All Is Lost*, *Oblivion* (2013), and *Gravity* (2013), punctuate the theoretical culmination of our investigation of earth and world. Each film presents lone protagonists struggling to regain a safe place to call home. In a sense they all represent an attempt to return to earth and find shelter there. In all three the world is gone and the effort to return home is met with a series of disorienting and traumatic events. Orientation toward the earth is paramount as our protagonists attempt their seemingly doomed missions to earth (or land, in the case of "our man"). These films can serve as a capstone to our exploration of ambivalence in relationship to both earth and world. These recent images from popular culture help frame an extended meditation on the trajectory we have been following in this book from Kant to Derrida and from politics to ethics.

Retracing our steps shows some of the limitations and possibilities that this route offers for developing an ethics of earth, or *earth ethics*, as a counterpoint to the dangers of technological and economic globalization. Revisiting the figures of island and desert from earlier chapters, only now as they show up in these films, highlights ambivalence in relation to earth and world. In conclusion *earth ethics* appears as a bridge and oasis, so to speak, that reconnects us to the earth and other earthlings and promises a way not only to survive but also to thrive. Earth ethics is a form of *terraphilia*, through which we love the earth enough to take responsibility for it. Here terraphilia becomes emblematic of an embrace of our ambivalent relations to earth and world. In the end, returning to earth via the Apollo missions to the moon with which we began, we explore the possibility of the saving power of global technology in relation to man-made climate change as the affirmation of our shared, yet singular, bond to the earth.

ALL IS LOST?

While Robert Redford's "our man" in *All Is Lost* is not literally returning to earth, his life depends on returning to dry land, to touching the earth. The film opens with the sound of water lapping, the image of a semisubmerged shipping container, and a voice-over saying, "I am sorry," "I tried to do the right thing, but failed," and "all is lost." We meet the enigmatic protagonist when he wakes up to discover that his sailboat is taking on water after running into the lost shipping container, which is leaking its contents, tennis shoes, into the sea. After a couple of stressful days, "our man" manages to patch the hole and bail the water, only to encounter a storm that rips off the mast and leaves his sailboat sinking. He takes refuge in an inflatable life raft with what he salvages from the sailboat, including an unopened sexton, apparently given to him as a gift (there is a card in the box with the sexton, but our man simply looks at the envelop and throws it aside, suggesting that either he has abandoned his past or it has abandoned him). Since all electronic equipment was damaged and rendered inoperable by the corrosive seawater, he is left navigating by the stars. In spite of his ingenuity, resourcefulness, preparedness, and well-kept gear, it becomes clear that this solitary man is no match for the forces of nature.

We never find out what he is doing sailing alone on a beautiful wooden sailboat (without a motor) in the middle of the Indian Ocean or who or what he has left behind. This man is an island, completely inscrutable and disconnected from any other. Our man is the maverick individual going it alone, autonomous and self-contained, not dependent upon anyone else. He is determined to survive his solo trip by virtue of his sheer determination and know-how. In the end, however, his insularity and insignificance make him invisible to the few container ships that pass him by without ever noticing his flares or his calls for help. It is as if his own self-contained self-mastery has been so successful that he is forever cut off from all others to the point of not being recognized by them.

The fact that his trouble begins with a shipping container full of tennis shoes, and that the giant ships that pass him by are loaded with the same kind of containers, we could interpret as commentary on globalization and man-made pollution (such that a huge container of cheap shoes is accidently dumped into the ocean, presumably without anyone noticing or caring). These ships of containers are moving massive quantities of goods thousands of miles because global commerce promotes outsourcing and the resulting consumption of vast natural resources. In addition, the fact that "our man" is sailing in a wooden boat without a motor, and that the little technology he has is ruined, suggests a struggle between a less technological age or mindset and contemporary global technology. "Our man" seems to prefer smaller, low-tech means of transportation that do not burn fossil fuels or pollute the oceans or atmosphere. And yet it becomes clear that he is no match for the forces of either nature or globalization; like his small wooden sailboat on the vast Indian Ocean, he is a thing of the past.

Our man resorts to his sexton and navigating by the stars, even though it is obvious that his fate is sealed. We know from the start that "all is lost," and still "our man" is determined to get his bearings, as if merely orienting himself will rescue him. He continues to mark his course on a map even though he is in the middle of the ocean and has no control over that course. It is as if the search for bearings, for orientation itself gives his voyage meaning even though it cannot save him. Finding himself adrift on the open sea is more bearable through his attempts to orient himself, even if that means discovering that "all is lost," which indeed he apparently denies until the very end. On the one hand, this futile activity gives meaning to his otherwise hopeless attempts and, on the other hand, it reasserts the fantasy of control in the face of his absolute lack of control.

In this way, "our man" represents another relic of the past, the autonomous, auto-mobile, maverick individual going it alone. In spite of his determination and his attempts to control his boat, he fails. Indeed, he is doomed to failure. He cannot survive isolated and alone. And, in the end, it is his failure to do so for which he apologizes. In other words, until the very end he maintains the illusion that he can succeed, that his survival depends upon himself alone. And he does not apologize for this arrogant attitude, or his hubris, but rather for failing. His last act is one of desperation, when he burns his life raft in an attempt to get the attention of a passing ship. The autoimmune logic of this gesture, whereby his survival becomes dependent upon destroying his own life raft, is telling once we take "our man" as a representative of an outmoded fantasy of man against nature, or manly individualism, which this film shows us is doomed to fail. And yet the film itself preserves this fantasy like a tiny ship preserved in a bottle. The film itself makes us reinvest in this fantasy. We are there with "our man" trying to control the elements, determined to survive against all odds. This fantasy of control and of the autonomous self-contained individual is still with us. It continues to fight for survival even as it is passed by without a backward glance amidst the "forward progress" of globalization.

Our man is the diehard individualist who refuses to admit defeat until the bitter end; importantly, he does not undergo any transformation. He is true to himself until the end. Indeed, it is only after we see him submerged, presumably dead, that he opens his eyes, sees a bright white light, and reaches for an outstretched hand. The film offers him this hand only after he has died. Only after the death of this isolated autonomous maverick individual is there hope of salvation. Only after he is swallowed up by the sea—Robinson Crusoe's worst nightmare—does our man make contact with something beyond himself. Again signaling the autoimmune logic of the fantasy of the autonomous self-contained and self-mastering individual, it is only after death that our man sees the light; through death he finally reaches out for what he has apparently denied throughout the film—or at least through setting sail alone in a small boat without a motor—that he needs anybody else. In the end his "salvation" depends on what he denies, something beyond himself. This last relic of liberal individualism opposing itself to both nature and technology is fated to die. And yet, like that hand reaching out at the end of the film, we want to hold onto it. Even as we see how tiny and puny our man is compared to the vastness of the ocean and the huge container ships that pass him by, we continue rooting for the little guy because we don't want to let go of the fantasy of mastery and control in the face of both nature and technology.

This brief meditation on *All Is Lost* reminds us that the ambivalence of the Apollo missions to the moon is still with us. Seeing our man floating on his round life raft alone in the immensity of the ocean conjures reactions not unlike seeing the earth floating alone in the immensity of space. Our man is rendered insignificant and small from the vantage point of the sea, especially in the shots from far below his life raft. Still, he is the hero of the story, which revolves around him, as he never gives up his attempt to control and conquer the sea. As the contradictory rhetoric of the Apollo missions made manifest, even while seeing the earth from space renders us small and insignificant, it also makes us the towering conquerors of both earth and space. Both these reactions obtain at once. We see ourselves as both the center of the universe and an insignificant speck, as both conquering space in our own name against the others and representing all of humankind in cosmopolitan unity. This ambivalent fantasy continues with our man, who represents every man's struggle against his own mortality and insignificance in the face of the vastness of the universe or of our own planet or even of the Indian Ocean.

The persistence of these ambivalent dreams of our importance and our insignificance, our self-contained isolationism and our desires for global cosmopolitanism, and our will to mastery and our acceptance of our own limitations becomes more apparent when we consider another recent film, in the theaters at the same time as *All Is Lost*, *Gravity*. For, unlike our man, the lone protagonist in *Gravity* survives against all odds and makes it back home to earth from space. And unlike our man, who is rendered small and insignificant at the end, "our woman" in *Gravity* is shown towering and huge, emerging from the murky water, like a monument grounded on earth looking up at the sky. At the end of this chapter, with *Gravity,* we will return to earth. In the meantime, the bulk of this chapter, we will retrace our steps across earth and world in order to contemplate an earth ethics grounded on loving the earth enough to take responsibility for it by embracing the fact that, even if we do not share a world(s), we do share the earth.

OUR ROUTE HOME

In the trajectory that we have been following from Kant through Arendt and Heidegger to Derrida, *Earth* is a limit concept and a concept of limit. While for Kant and Arendt this limit is an end point, for Heidegger it is the beginning of an alternative way of relating to the world. Kant argued that the limited surface of the earth obligates each one of us to welcome visitors; based on this physical limitation, the right of hospitality directs that we share the earth. In spite of the shortcomings of Kant's analysis— that ultimately sharing the limited surface of the earth is about commerce and private property and that nonhuman living beings and other nonhuman earthlings are considered resources for human use—the notion that human beings are a species that evolves, particularly in terms of new possibilities of political and ethical commitments, is promising. Kant's philosophy suggests that we are evolving to be a more ethical species. And while Kant would not consider this possible for nonhuman animals, developments in zoology and biology suggest otherwise. Already with Darwin the theory of evolution was not just about the survival of the fittest but also the importance of cooperation and compassion among and between animal species. Admitting that we are an animal species that evolves—for better or worse—is already a step toward an ecological ethics that takes into account our place within the biosphere, which we share with all other living beings.

Like Kant, Arendt's theory of politics is also based on sharing the earth. But for Arendt it is human dependence and plurality rather than human reason that obligates us to share. No one of us can live in isolation. Rather we share our existence, which is much more than sharing physical proximity or physical space; it is sharing meaning through traditions and cultures as well as embodied interactions. Our plural existence is based on sharing a world and worlds, the totality of which makes up *the* world. For Arendt, cohabitation is part of the human condition, and denying it not only leaves us worldless but also refuses our humanity. Although Arendt discusses our relationship to things in the world, for her these relations are mediated by the meaningful political worlds of relationships between human beings. Like Kant, Arendt does not consider nonhuman animals as capable of politics or of having, or making, a world. For her, our very humanity is at stake in the plurality of worlds that result from cohabiting the earth. She does not consider other animals or other living beings as cohabiting the earth with us. Rather, like Kant, other living beings are reduced to mere resources for human labor and work, which is to say for human reproduction and production. So while the human world is dependent upon these other beings, they do not cohabit the earth, nor do they contribute to the plurality of worlds, which Arendt insists is a necessary part of the human condition. She does not consider—at least directly—the ways in which the human is conditioned by and through its co*habitation* with other animals and other living and nonliving beings. And while she insists that human beings are earthlings through and through, she does not acknowledge that we are not the only earthlings—not by a long shot. As far as earthlings go, we are vastly outnumbered. Furthermore, unlike those telling moments when Kant embraces the human as a developing species living among other species, Arendt rejects reducing the human to its species. Her worry about an alien seeing cars from space and concluding that human beings are giant snails or turtles with shells belies the value—or lack thereof—she puts on these other ways of being in the world. She is critical of human rights in the abstract precisely because the notion of the *human* in itself does not capture the uniqueness of our place on earth, namely as inhabiting the world and as giving ourselves rights. We might say that, in the name of humanity, she rejects the bare notion of the human. Our humanity, for Arendt, cannot be reduced to the fact that we are members of the human species. In order to contribute to earth ethics, Arendt's notion of shared cohabitation must be stretched to nonhuman animals, other living beings, and to the earth itself. In other words, cohabitation must be extended to include all earthlings and the earth.

Heidegger, too, emphasizes the unique place of the human in the world, first as world formers and later as caretakers. He is interested not in the political world, but rather the ontological world of relationships that grounds all relations. For Heidegger, inhabiting the earth is much more than sharing its surface or cohabiting with other people; it is rooted in the *fourfold* of earth and sky, mortal and divinities, all of which signal something beyond human mastery or control and all of which make up *the world*. The fourfold as *das Geviert* suggests boundaries and limits that allow each element to belong with the others. Each element belongs to, and limits, the others. And the boundedness of the four opens onto the world. The limitations of the earth, the earth's refusal and self-seclusion, open onto our world. And, although Heidegger shies away from politics and ethics, we could say that these limitations are not the end point of politics or ethics, but rather the beginning of our being together in the world. Each living being—each being—comes into being, and into its own, through its relations with its environment and the creatures with whom it shares that environment.

Although Heidegger focuses on Dasein's unique being in the world and Dasein's unique participation in the fourfold, we must insist that the primary relationality of *each being* not only gives it a sense of belonging but also moves it outside of and beyond itself. The world is the constant fluid flowing event of beings coming into being through their relationships to earth, sky, mortality, and divinity. In more familiar terms, we might say that the world is the constant coming into being of beings through their relationships to earth and its atmosphere. Earth's biosphere is made up of finite creatures that are essentially bounded, most primarily by birth and death, and yet whose relationships transcend that finitude and always point beyond any individual life. Dasein is not the only way of being in this relationship to finitude; Dasein is not the only mortal limited by death. Indeed, as Heidegger's work reminds us, we cannot know the animal's way of being in the world, or any particular animal's way of being in the world, because of their "telling refusal"; they refuse to tell us about it. Other modes of being in the world, other ways of having or not having world, are limits to our own understanding of what it means to have a world. Furthermore, given Heidegger's introduction of the fourfold of the event of being in the world, we must consider that not only nonhuman animals but also other living and nonliving beings necessarily participate in this event, or this flowing experience that makes up the world, always in relationship and always limited by the earth and bounded by the other elements.

Leaving behind Heidegger's fourfold, we could simply say that the earth is a limit to world. First and foremost, every world is grounded on the earth. As different and various as they may be, each world depends upon the earth. And every world is born—or hatched—from it. What a world can become is both engendered, and limited, by its essential connection to earth. Whereas, for both Arendt and Heidegger, the *world* is uniquely human, all creatures share the *earth*. In Arendtian terms, the human condition is governed, if not determined, by our relationship to the earth. Furthermore, as we have seen, the earth operates as a limit concept in the philosophies of Kant, Arendt, and Heidegger. The earth, then, not only literally determines what we are and what we can do but also figuratively reminds of our limitations. In these works, the figure of earth reminds us of our own limitations insofar as we are not the masters of either earth or world. Earth ethics, then, is an ethics of limits, particularly learning to limit our own hubris and will to mastery.

Although earth is not a central concept in Derrida's writings, he does call his project a "philosophy of limits" insofar as he is attempting to multiply limits (Derrida 2008). In other words, he challenges acknowledged limits in order to open up new frontiers by "deconstructing" accepted oppositional binaries (cf. Derrida 2009, 15–16).[1] Of interest here is the way in which Derrida challenges the boundary between ethics and politics, particularly with his notion of justice. In Derrida's latest work, Kant and Arendt's concern for the "we" of politics becomes a concern for the "I" and "you" of ethics. Still moved by the fundamental relationality that grounds our experience, Derrida questions whether *we* truly share a world in the sense of a political *we*, even if this *we* is an Arendtian plurality rather than a Kantian universal. Opening the anthropocentric and humanist worlds of Kant, Arendt, and Heidegger onto nonhuman others, Derrida asks whether we share the world with animals and whether they share a world. Exploiting the tension between the political *we* and the ethical *you*, between sharing a world and the singularity of each living being, Derrida's latest work suggests a more ethical polis, what in earlier work he has called *justice* or *justice to come*.[2] Indeed, Derrida suggests that the tension between ethics and politics is necessary for justice to come. In a sense the gap between the two is the space that holds open the impossible possibility of justice. Insofar as ethics' demand for singularity and politics' demand for equality cannot be reconciled, justice is always deferred, which makes vigilance all the more necessary and urgent.

If, as Derrida maintains, sovereignty is haunted by an autoimmune logic that destroys what it purportedly protects, then both the political subject, "we," and the moral subject, "I," must be rethought. The political subject and the moral subject can no longer sustain the fantasies of wholeness, autonomy, and sovereignty assigned to them by traditional political or moral philosophy. They cannot be imagined as autonomous self-governed islands that exist in isolation. For his part, Derrida invokes the figure of the island to conjure the radical singularity of each living being, of each life, and perhaps each moment of each life. He does so by imagining the ethical relation with its absolute, and yet impossible, responsibility when the "world is gone" and one becomes responsible for the other, to carry the other. Derrida creates an ethics of urgency—we might even say ethics as trauma—that engulfs and overtakes any self-same, self-certain, autonomous sovereignty. Derrida's figuration of the singularity of each as an island, however, and the resulting hyperbolic responsibility to the other when the world is gone, does not seem to lend itself to a sense of belonging or community. His unworldly ethics does not provide an ethos of earth that takes us far enough beyond the model of Kantian hospitality, even if now unconditional hospitality, and toward a sustainable ethics based on our singular relation to the earth, a relation we also share with other earthlings through the "miracle" of cohabitation.

If for Kant the world is constituted in terms of the finite surface of the earth, while for Arendt the world is constituted in terms of the plurality of worlds formed through human relationships, and for Heidegger the world is the shifting event of the fourfold relationality between earth, sky, mortals, and divinity, for Derrida the world is the singularity of each life. We have moved from world as universal to world as singular event. In this way, we have moved from politics, with its concern for universal rights, to ethics, with its concern for particularity; that is to say, from politics concerned with equality for all to ethics concerned with the singularity of each. And we have moved from a philosophy focused solely on human life to a philosophy affected by all life, every life, a philosophy that leaves no stone unturned. But, we might ask, how can we return from this place? How do we get back to the world from the face-to face obligation to the singular other? How do we return to the world once the world is gone? As Kierkegaard might say, it is not the leaping, but the landing that is difficult. For Derrida, this is the difficulty of justice for which we are responsible even if its achievement is impossible. We must act for justice even if every action risks injustice. This is the difficulty of bringing ethical obligations to bear on political action. And, in the case of earth ethics, it is a case of the world being grounded on earth. When the world is gone, perhaps we can still find our way back to earth.

How do we bring the urgency of the ethical obligation to the other, to each other, back into the call for cosmopolitan politics? We learn from Derrida that both individual and state sovereignty revolve around an autoimmune logic through which every attempt at securing and protecting borders also destroys them. If this is the case, then nationalism can never be self-sustaining; it must necessarily give way to cosmopolitanism. And, yet, absolute cosmopolitanism, with its hyperbolic demand for unconditional, rather than Kantian conditional, hospitality is impossible. Furthermore, the sovereignty of both national *law* and cosmopolitan *law* suffer that same fate, at least in terms of deconstructive logics. Neither can maintain the sovereign authority necessary for protecting rights or enforcing responsibilities. This is where Derrida inserts poetry, not only the repeated poetic fragments that work to dislodge philosophical categories but also the call for a poetic alternative to sovereignty, a poetic sovereignty, which is to say one that undoes itself and in so doing creates an alternative world, a prosthetic or surrogate world, to support the other when the world is gone. Unlike the prosthesis of ego or the apparatuses of state control, however, poetic majesty supports the world not by claiming that it is the most, the best, or the absolute, but rather, to the contrary, by showing us that there is always more in excess of what we see, or think that we see, there. Furthermore, rather than prostheses or stabilizing apparatuses through which we attempt to create the scaffolding of mastery or certitude, poetic structures elude mastery and escape certitude. The way in which poetry or poetic majesty supports the world is not the way in which political or individual sovereignty does so. Neither is it the way in which the things of the world, the products of work, do so for Arendt. The stability provided by poetic sovereignty is an unstable stabilizing force. It is a groundless ground, which for Heidegger always brings us back to earth.

We could say, like Heidegger's conception of earth in "On the Origins of the Work of Art," poetry *juts through* the world. This poetic world-making, sometimes literally in the form of a poem, forges addressability and belonging that are the building blocks of both subjectivity and community. The bond forged through poetic world-making is not that of sovereign authority that compels through the force of law or military might, but rather through the sense of belonging, being heard and hearing the call of the other. Poetic world-making is not the world-forming that Heidegger claims is the privilege of man alone. Rather, following Derrida, especially when he follows the animal, we see traces of this poetic majesty in every thing that leaves a trace; this broad notion of the poetic opens onto the mysterious and excessive worlds of interpretation, whether interpretation by human biologists, zoologists, geologists, or artists, or interpretation by nonhuman animals as they interact with other species and their environment. Every being that is responsive is capable of poetic majesty. Or, with a more Derridean twist, we might say that every being that is responsive *suffers* poetry. The dynamic of poetic sovereignty that opens up, rather than closes off, the possibility of response is akin to what elsewhere I call *witnessing*, or the structure of address and response. In its own ways, every living being addresses others and its environment, and every living being is responsive to them. Even when our response is to eat them, awareness of our dependence on them and our profound relationship to them, can make a difference in our attitudes toward them and our treatment of them. Awareness that we are not their sovereign masters, but rather fellow creatures inhabiting the same planet can help us rethink the fundamental questions of this project: Can we share the earth even if we do not share a world? Can we expand the horizons of poetic attunement to the earth, and to the multitude of different worlds that coexist on it, by acknowledging that each species and each singular living being together belong on this earth, our shared home? Can the vantage point of the earth give us new ways to view relations between different worlds, what it means to share *a* world or *the* world, and, furthermore, what it means that every world is of the earth?

The limited surface of the earth does not in itself guarantee sharing with others. Indeed, as we know, all too often, the limited resources of the earth lead not to sharing, but rather to hording and violence, even war. Perhaps the law of hospitality can be extracted from the limited surface of the earth and extended into an ethics of earth. Maybe there is an ethical dimension to what Kant calls our *common possession* of the earth, but not if we imagine the earth only in terms of its divisible surface.[3] And not if we conceive of individuals as interchangeable occupants who could just as easily be in one place as another, as Kant does. What if we complicate our notion of earth and conceive of it as a Husserlian basis body or primordial arc? If we realize that we are earthlings who cannot occupy any other place, this may add to the ethical dimension of the Kantian insight that we must live together on the limited surface of the earth. While our conception of the earth and our relation to it has gone beyond the mere ground on which to stand, now it has become an abstract pregiven against which, and through which, we measure all our experience toward an ethics of earth. If its normative force is also pregiven, then we can never know or understand it. And perhaps this is the case. Still, the Husserlian realization in itself does not obligate us to share or to care for the earth or for our fellow earthlings.[4]

Certainly, Arendt's claims that plurality is the law of the earth, and that cohabitation is part of the human condition, go further than Kant's common possession of the limited surface of the earth or Husserl's earth as the primordial arc for all of our experience. For, in addition to determining and limiting our political and economic exchanges à la Kant, and forming the pregiven basis for all our knowledge and understanding à la Husserl, the Arendtian earth survives and thrives through plurality and diversity; we could say that the earth survives and thrives through a plurality of worlds. Extending Arendt's notion of natality beyond the human world, however, it is not just human life that is conditioned by natality (which ensures that every moment someone new is born) but also potential worlds are created anew with every living thing that is born, spawned, or hatched, perhaps even germinated or sprouted. Each one is unique and each one changes its world and the world around it. Plurality is the law of the earth; and the earth is inhabited by a plurality of worlds, past, present, and future. No longer a fixed and finite surface that can be mapped and distributed, nor an abstract principle of pregivenness, the earth has become a dynamic living biosphere inhabited by human beings who are necessarily social and political beings and necessarily dependent upon the earth and each other. Moving away from Arendt, no longer a merely human world or human-centered world, earth ethics necessarily moves us beyond ourselves and toward other earthlings with whom we share the planet. Our interconnectivity extends far beyond the human world or human worlds.

We have moved from the Kantian notion of the earth as divisible surface, through Arendt's earth as plurality of worlds, to Heidegger's thicker notion of earth as a telling refusal in relation to ways of being in the world. Heidegger's insistence on separating earth and world makes clear the ways in which the earth acts as a limit to world. More specifically, the earth limits the human world and human world-making. The earth's refusal tells us something essential about our world and our life on earth. One thing that it tells us is that the earth cannot be reduced to its surface or circumference, nor can it be reduced to elements or even to its physical attributes, particularly not in terms of measurements, quantities, exchange values, or property deeds. Furthermore, the Earth is not just one planet among others. Rather, the earth is our home; and it is home to life as we know it. Thus the earth is a boundary for all our experience. Participating in Heidegger's fourfold, the earth is limit and excess, the beginning, but never the end of meaning, the event of being. No longer a divisible surface, earth is an indivisible way of being, our relationship to which determines the meaning of our world(s) and the quality of our lives.

This thicker conception of earth also gives us a thicker conception of ourselves. As our idea of earth becomes more complicated, so too does our idea of ourselves. Just as the earth is no longer imagined in terms of a calculable and mappable surface, human beings are no longer imagined in terms of calculable and mappable categories of understanding. Human relationality and plurality give rise to unpredictability that foils rational calculation and, in some sense, makes it beside the point. Just as the thicker conception of earth gives us a thicker conception of ourselves, so, too, it gives us a thicker conception of others, especially nonhuman others. This thickening suggests that we cannot draw set boundaries around species or organisms, peoples or individuals.[5] Rather, the boundaries are fluid and dynamic, and limits are constantly multiplying and exceeded. And, yet, we survive and thrive by virtue of our singular bond to the biosphere of earth. Human being's bonds to other creatures and our singular bond to the earth begin to suggest an ethical dimension that the limited surface of the earth does not.

Thus it is not the limited surface of the earth alone that obligates us to others.[6] Nor is it the mere fact that we all inhabit the same planet. Physical proximity alone does not obligate us to each other and to the earth. The law of plurality, which demands diversity of worlds, does not go far enough to safeguard the singularity of each or our singular bond to the earth. Even the ethical obligation that results from the face-to-face relation with a singular other does not necessarily entail responsibility to the earth and all its inhabitants. Rather, it is our singular bond to this planet, a bond we all share, that makes us responsible and able to respond, or response-able. The biosphere is made up of responsive beings, in a complicated cohabitation that makes life on earth possible. Without the earth and our fellow earthlings, we could not survive, let alone live meaningful lives. The meaning of our lives, what makes our habitation a *world* and what makes the earth a *home*, is born out of our cohabitation or, more precisely, out of our bonds with the earth insofar as we are singularly dependent on it. We cannot live elsewhere. And therefore we are responsible to and for that by virtue of which we live. Our common cohabitation alone, especially as articulated by Kant in terms of geography and earth's divisible surface, does not affirm the singularity necessary for the radical ethical bond.[7] But our unique bond to the earth, and the fact that we *belong* to the earth, may be the basis for an ethical imperative that goes beyond either the Kantian universal based on the earth's limited surface or the Derridean singular based on the disappearance of world.

The singularity of the bond to earth that is shared by all earthlings is what turns this proximity of all, and singularity of each, toward an ethics of earth. The uncanny singularity of a shared cohabitation on the same planet, our one and only home, brings with it responsibility. While each species and each individual member of that species has its own world, which determines its relation to earth, all share the fact of radical dependence upon this one and only earth. The singularity of our bond to the earth demands plurality and diversity and yet it is something that we share, in our radically different ways. This plurality is not just in terms of numbers or quantities of worlds or lifestyles, but rather relationships and coevolutions that make up a balanced ecosystem, perhaps recalling Kant's notion of dynamic equilibrium. All earthlings depend upon biodiversity and ecosystems to live and thrive. The earth is a dynamic of relationships that cannot be reduced to its mappable surface or to the astronomical notion of a planet as a heavenly body. Rather, each singular being is part of an ecosystem that ties it to the earth and to other earthly creatures. The earth is a *dynamic living network* of relationships that supports life and death. The earth supports the mortality of all living creatures. Whether or not we share a world, we do share the earth.

Thus earthly responsibility obtains even if we do not share a world. For Derrida, ethics begins there, where the world recedes, in the encounter with the other that is so foreign and different it is beyond recognition. This is when decisions become difficult, namely when traditions and familiar frameworks abandon us. And, still, we still must decide and act. Ethics begins when the world as we know it ends. The limits of world open onto the beginnings of ethics. Earth juts through world. Beyond tradition and laws, beyond reason and recognition, we are responsible. What Derrida calls *hyperbolic ethics* takes us beyond the notion of *ethos* as cultural mores with which we are familiar. Hyperbolic ethics makes this ethos uncanny insofar as we cannot be sure what we are responsible for, only that we are responsible. Derrida sometimes describes this radical responsibility as *impossible* and yet binding nonetheless, even though it is beyond our control and mastery or precisely because it is beyond our control and mastery. For, if it was in our control, and we could master it, ethical responsibility would be calculable, predictable, and mechanical.

We are in the uncanny position of being responsible for what we do not know, cannot predict, and never control. Uncanniness is the ethos of philosophy, which makes the familiar appear strange, as if seeing it for the first time.[8] And uncanniness is the ethos of our earthly home. Earth ethics is an uncanny ethics based on a shared singular bond that juts through the plurality of worlds. Within earth ethics, earth and world, or earthly and worldly, so long disavowed and denigrated by Christian traditions as sinful and fallen, are not redeemed or saved, but rather cherished and nurtured as the uncanny places we call home. Indeed, earth ethics is grounded on the ambivalence of this uncanny place between earth and world, along with the acknowledgment and embrace of that ambivalence. Earth ethics is an ethics of ambivalence insofar as it revolves around tension between earth and world. Or, in other terms, we could say that earth ethics revolves around radically singular worlds not only necessarily coexisting on the same planet but also sharing a singular bond to that planet, our uncommon common earthly home.

If, as Heidegger urges us, we question the traditions and values we inherit in order to build our world, then, taken to the extreme, as Derrida suggests, when that world is gone, its traditions and norms are gone, and all that is left is our responsibility to the other, whether this other is another human or nonhuman being. Extending Heidegger's thought, we could say that what is left when the world recedes is our place on the earth in relation to the sky, other mortals (now including nonhuman animals), and divinities or simply something beyond our mastery and control, something essentially mysterious, including meaning itself. Even when the world is gone, the earth remains. We might say that the excess of the singularity of each is the excess of its relationships and relationality, which are all mediated by the primary relationship to the earth and the biosphere. What is more, divinity dwells in this space in between beings. The excess of an encounter between individual beings or between species takes us beyond ourselves and, at the same time, reminds us of our limitations. For, in addition to the radical singularity of each is the radical dependence of each, not only on the others, but also on the earth. When there is a balance of ecosystem, diverse species coexist and thrive; but when the ecosystem is out of balance, one species may wipe out another. And yet life goes on. This is not so with the earth itself. Life cannot continue if the earth itself is wiped out. The bond to the earth is fundamental and shared by all. Every life-form participates in an ecosystem upon which it is dependent for survival and to thrive, and every ecosystem is inherently tied to the earth. The earth shelters each singular individual being and every species. Avowing, and then embracing, this interdependence and fundamental relationality is the beginning of earth ethics. For, without it, we suffer under the illusion that the singularity of each means that every man—or every being—is an island. Encounters through difference remind us of our own limitations and the ways in which the earth's refusal, or other animals' refusal, makes us *as if* we were islands where creative or poetic understanding from the heart may be the only possible bridge.

While Derrida takes us to this extreme of every being is an isolated island in order to emphasize the radical singularity of each, he also insists upon the fact that we share the world not just with other human beings but also with all creatures and perhaps even with stones.[9] Derrida is concerned with "every world, no less," or no less than every world. For the only way that politics can avoid becoming totalizing and homogenizing is with the constant interruption of the singular and the singularity of each. This singularity is the counterpoint to the universality that Kant proposes and that Arendt and Heidegger warn against. And yet it is not only the leap out of the world and toward the singular other that defines ethical movement but also the landing back into the world, our own world, in ways that transform that world by opening up the response of those others. This gives us a new conception of response-ability as an obligation to open oneself up to the other's response and, furthermore, to act in such a way as to open up the possibility of the other's response without ever knowing with certainty what this means. This notion of responsibility allows the ethical to jut into the political, the earth into the world. The impossible situation of responsibility is the result of this encounter between politics and ethics, world and earth. Politics demands equality for all, while ethics demands the singularity of each. Politics requires rules and laws that apply to everyone, while ethics is always in excess of rules and laws because it requires a decision beyond mere rule following. Politics must be calculable and generalizable; ethics must be incalculable and singular. Our responsibility lies in this strife between politics and ethics or what we could call the strife between world and earth.

An ethics of earth, or earth ethics, is one in which we acknowledge and affirm our singular bond to the earth and its biosphere. The fact of cohabitation on earth must become the ethical decision to affirm it. We must affirm our common bond to the earth, commit to hospitality, and avow our radical dependence on others and on the environment. We can take up our common cohabitation on earth as the ethical and political ground of relations with others, including nonhuman others. Doing so requires not only avowing our dependence on others and the earth, by virtue of which we exist, but also committing ourselves to interpret and reinterpret, to continually question, the meaning of those relationships. We are bound to others on the earth—that much is given. But what we make of that bond is what becomes significant for ethics and politics. We must take up the question of what it means to dwell on earth and to cohabit the planet with other earthlings. That bond can become the basis of an ethics of the earth that obligates us to others with whom we cohabit and obligates us to the earth on which we live as our singular uncanny home.

LOVING THE EARTH

This brings us back to the series of questions that motivate our explorations of earth and world: Can we formulate an ethos of the earth that is not totalizing and homogenizing? Can we conceive of an earth ethics beyond the globalization of technology and beyond global markets that make everything fungible and disposable? Can rethinking earth help us to reconceive of the planetary and the global in ways that open up rather than close off ethical and political responsibility? Moving away from Kantian universalism, and yet maintaining his vision of cosmopolitan peace grounded on the earth, through Arendt and Heidegger's warnings against the dangers of global thinking and global technology, to Derrida's meditations of singularity of each life as *the world*, can we reconceive of our relation to the earth and to each living being as nourishing and sustaining both life and thought? Echoing Arendt, what would it mean to learn to love the world *and* the earth enough to take responsibility for them? It would mean loving diversity and plurality of worlds and embracing the refusal of other beings to be mastered and controlled by our understanding. It would mean loving what we cannot know, the mystery and uncanny thrill of being alive among so much life.

At the end of our exploration of earth and world, we can reformulate the questions: What does it mean to *love* the earth enough to take responsibility for it? What is the relationship between love and responsibility? Certainly, an ethical notion of love cannot be reduced to stereotypes or fantasies of romantic or familial love. On the one hand, an ethics of love is an acknowledgment of the role of affects in ethics. Affects are the force that binds us to our obligations and compels us to act on them. Affects involve not only emotions but also sensations. Love of the earth or *terraphilia*, then, is an emotional and sensual connection that binds us to others and to our environment. On the other hand, like cohabitation, love is unchosen. And yet, to become ethical, it must be chosen and affirmed. In this sense, love is the affirmation of our bond to others and to the earth. To become ethical, and to ground political commitments, our dependence on others and on the earth must be avowed and then taken up, poetically or responsibly in the sense of opening onto, rather than closing off, possible interpretations, possible worlds.

Cohabitation is unbidden, and yet we can affirm it. With Augustine, we can say, "I want you to be." Or, with Martin Luther King, we can say, "I decided to love." Or, without saying those words, perhaps without language, we can act in such a way as to affirm the other's existence. Love, like cohabitation, is an essential part of the human condition. But it is also an essential part of the animal condition if we consider that most animals rely on cooperation and empathy to survive and thrive. In this sense love is an emotional sensual bond and more; it is also a life force that bonds individuals to each other and to the earth. It is the connective tissue that binds each living being to others and to their environment. Obviously, this is a much different, and broader, notion of love than our everyday usage, or the romantic notion, of love, particularly insofar as the ethical responsibility to love, or the love that makes us responsible, paradoxically, requires that we choose the unchosen. We must cooperate and share in order to survive and thrive and yet we must also affirm cohabitation and the existence of each and all.

It is not just our own belonging to earth, then, that is at stake in earth ethics. Rather, all earthlings belong to earth. And the earth belongs to earthlings. In terms of earth ethics, this belonging requires an affirmation of the sort Augustine names when he says, "I want you to be." This affirmation of the other's existence can foster a sense of belonging. We must be careful, however, not to fall back into the logic of recognition in which the other's being is conferred by those in power, where the existence of the colonized or dominated requires the affirmation of the colonizer or dominator. For this notion of recognition assumes the kind of sovereignty and mastery questioned throughout these chapters. Rather than conferring belonging on another through a sovereign affirmation, this notion of the affirmation brings with it an ethical obligation to free other beings into their own belonging to their environments and to the earth. Indeed, acknowledging that all living beings belong to the earth is the beginning of such an affirmation. Furthermore, the realization that the earth belongs to us as our shared home through our cohabitation with others, rather than as our possession or a resource that we own, grounds an ethics of earth as an ethics of responsibility to care for that home. This earth ethics is necessary to ground a politics that is responsive to the needs of other earthlings.

Love becomes the basis for political action when this ethical obligation is put into practice through civil law and public policy. This points to the affective and bodily dimension of cosmopolitanism driven by an ethics of earth. Of course, in practical terms, the unconditional principles of hospitality and affirmation of each and all must be tempered by the need to act and to navigate our world in proximity to others. Indeed, negotiating practical political concerns may require that we do not extend hospitality to all, for example, Eichmann, or others who do not respect plurality or diversity, or those who brutally close off the possibility of the others' response, may not be welcomed. And yet we are obligated to hold open that possibility. In this way, the force of the ethical—of loving enough to take responsibility—must jut through political practice.

Love provides the force of the ethical and political obligations to others (human and nonhuman) and to the earth (organic and inorganic) through the realization of our fundamental dependence upon them. In this way, love of earth is not just altruism toward other living beings and the ecosystems that support them but also a drive toward our own survival as creatures deeply embedded in those ecosystems with those others. As some biologists argue, we have evolved to share and care because those values are in our own best interests as well as the interest of the planet.[10] Today, as we contemplate man-made climate change and man-made pollution, however, we have to wonder whether our industrial evolution is putting the earth at risk. Ultimately, caring for the earth, acting as its steward, may involve moving beyond concerns for human survival and toward the survival of the planet along with its billions of nonhuman inhabitants.

Indeed, even when we imagine ourselves as the stewards of earth, like Kant's lice who imagine themselves inhabiting a world and an earth while living on the mere head of another animal, we risk taking ourselves to be the alpha and omega of existence. When we take ourselves to be the center of the universe, like Kant's lice we become self-aggrandizing parasites on earth:

> Because in its [lice] imagination its existence matters infinitely to nature, it considers the whole of the rest of creation as in vain as far as it does not

have its species as a precise goal, as the centre point of its purposes. The human being, so infinitely removed from the highest stage of beings is so bold as to allow himself, in a similar delusion, to be flattered by the necessity of his existence. The infinity of creation encompasses in itself, with equal necessity, all natures that its overwhelming wealth produces. From the most sublime class among thinking beings to the most despised insect, not one link is indifferent to it; and not one can be absent without the beauty of the whole, which exists in their interrelationship, being interrupted by it.

(2012A, 296-297, 1:353-354)

At this point we might ask, even if it is in their own best interests, do parasites have an *obligation* not to kill their host?

Even Kant saw that the earth is like a living organism that depends on complex interactions between its parts, or "organs," for its "beauty" and for its very survival. Some biologists, most notably James Lovelock, maintain that the earth is a self-regulating living organism made up of other living organisms in interlocking relations with rocks, oceans, and atmosphere. These interlocking relationships must be maintained in a delicate balance because if any one part comes to dominate, the whole is threatened. The most obvious example is the human dominance of the planet, which is threatening the whole with dangerous levels of pollution. This leads Lovelock to argue, "it is the health of the planet that matters, not that of some individual species of organisms" (1988, xix). And yet the health of the planet is intimately tied to, if not the consequence of, interlocking individual species of organisms. Even so, considering the health of the earth, rather than the survival of our own human species, gives us a different perspective on what it means to "save the earth." The earth and its biosphere, however, are inseparable. For what is the earth without it? It would be just another lifeless planet. As inhabitants of earth's biosphere, we—that is, all living creatures—share the fact that we cannot live anywhere else. And while in itself this fact is not a prescription for ethical or political action, it should play an essential role in the development of our ethical and political norms. Indeed, some biologists maintain that *biophilia* is not only essential to life but also a prime motivator for human behavior. For example, Edward O. Wilson argues that biophilia "is the innately emotional affiliation of human beings to other living organisms" (1993, 31). And extending Wilson's biophilia hypothesis, Stephen Kellert claims that human values and human fulfillment are inherently and profoundly tied to our relationship with nature and with other living beings on earth (1993, 42–43). On this view, human values and human development are a result not only of our relationships with other life forms but also of our love for and affiliation with them. Biophilia, they argue, has distinct evolutionary advantages for individual species, interspecies relationships, and ultimately for the biosphere itself. It is impossible, however, to conceive of the biophilia hypothesis without also imaging a terraphilia hypothesis. For everything we know of life is supported by, and exists upon, the earth. To love other living creatures and "lifelike processes" is to love the earth *and* its biosphere (cf. Kellert 1993, 42). After all, the biosphere *belongs* to the earth.

In this regard, the Apollo missions that transmitted the first photographs of earth from space were a moment of terraphilia coming to consciousness or, we could say, the self-consciousness of terraphilia. Astronauts and earthbound alike marveled at the beauty of the "blue marble," which was unlike anything else in space. Perhaps this uncanny moment made us aware that we share the earth as our home. Through this appreciation for our earthly home, the Apollo missions to the moon turned us back toward the earth. And whether the reaction was terraphilia or terraphobia, our dependence on the "pale blue dot" became achingly evident.

EARTH ETHICS

How can we develop an earth ethics based on a terraphilia hypothesis? In other words, how can we develop a notion of responsibility to the earth based on love of it? In order to answer these questions, we need to clarify what we mean by *earth*, *responsibility* and *love*. This entire book has been about what we mean by *earth*. And, as we have seen, within the history of philosophy, most especially with Kant and Arendt, the concept of earth is linked to political responsibility. Now their insights are extended into the realm of ethical responsibility by turning to Heidegger's conception of earth and Derrida's conception of world. What becomes clear is that by Earth we mean much more than a planet among other planets. Earth is the home to all living creatures (at least as far as we know). The earth is a network of relationships and connections that include its complicated biosphere, which is dependent upon its lithosphere, hydrosphere, and atmosphere. Indeed, we could propose that the biosphere, lithosphere, hydrosphere, and atmosphere are a type of Heideggerian fourfold insofar as they take on meaning in our worlds—that is to say, the worlds of earthlings. What we mean by *earth*, then, is a rich and complex relationality that sustains all earthlings, organic and not, in our shared *home*. More about *home* later, but for now it is important to remember that *our shared home* connotes belonging because *we belong to the earth* and not because the earth belongs to us, at least not as our possession.

What does it mean to love this earth, this rich network of relationality that sustains earthlings as our shared home? At this point, it is instructive to consider that the ancient Greeks had several words for different kinds of *love*. *Philia* is usually associated with Aristotle's discussion of friendship, which has many different forms. At its best, *philia* involves affection and fondness for another or others, along with altruistic actions that benefit those others without concern for self. *Agape* is associated with spiritual or unconditional love and becomes a centerpiece of the Christian New Testament. This type of love is selfless and, at the extreme, even sacrificial. *Eros* means erotic or intimate love, passion, and longing. For Plato, however, this longing is associated with creativity, such that *eros* gives birth to the highest forms of thought.

All of these forms of love contribute to love of the earth and of other earthlings. Our literal and figurative kinship with other species, and our dependence on them, may be conceived in terms of *philia*. Caring for the earth through its "telling refusal," which always points beyond our mortal existence, may be conceived in terms of *agape*. Indeed, caring beyond our own selfish needs may require *agape*. And *eros,* as love and longing, gives rise to our greatest creativity and contemplation through the strife of uncanny encounters with otherness. We could say that the "miracle" of cohabitation across vast differences, differences so great that we cannot even begin to understand each other, gives rise to creativity and contemplation. Creativity and contemplation, then, are not the result of sovereignty, autonomy, or mastery but rather of dependence, belonging, and deeply shared bonds with those whom we may not even know exist. For Arendt, the creativity born from love offers an oasis in the desert of meaninglessness that results from denying the plurality and natality of our existence. Creativity and contemplation are the result of unpredictability and not mastery, the uncanny unpredictability that is characteristic of our shared home, planet Earth.

Sigmund Freud links love with *eros* and *eros* with life. This connection between love and life may provide a starting place for thinking about love of earth. In his early work Freud opposes eros to ego as the drive that connects us to others and thereby keeps us alive both as individuals and as a species. Ego, on the other hand, is what separates us off as individuals and puts us at odds with others. Eventually, Freud formulates *eros* as the counterbalance to *thanatos*. Again, eros is the drive for life that connects us to others and the world while thanatos is the death drive that longs for equilibrium. In this regard, eros is dynamic and longs for relationships with all of their tensions and unpredictability, while thanatos wants stability and longs to overcome all tensions and return to a steady state. The death drive is the desire to avoid all tensions, even the pleasurable ones. Eros, on the other hand, is love as strife. Eros, then, is the dynamic life force that binds us to others. In this sense, it is from eros that we get compassion and tenderness, along with passion and erotic love. In Darwinian terms, eros is the social instinct that drives all sentient beings toward tenderness, compassion, and cooperation. In fact, Darwin imagines the evolution of tenderness and "sympathy," which become "virtues" that are passed on, initially by a few, until they spread and eventually become "incorporated" into life as we know it (1981 [1871], 101). Eros not only gives rise to compassion, cooperation, and sympathy but also to empathy and play. In other words, social bonds are formed through various manifestations of eros as the dynamic force of life.

Zoologists and primatologists have confirmed that play is important in establishing empathy and social bonding in many animal species, including humans.[11] For example, recently, psychologist Alison Gopnik proposed that "humans' extended period of imaginative play, along with the traits it develops, has helped select for the big brain and rich neural networks that characterize Homo sapiens" (Dobbs 2013). And neuroscientist Paul MacLean argues that play is essential in the evolution of empathy in the human species. Moreover, he links play to the formation of a sense of social responsibility (MacLean 1990, 380). There is increasing evidence that empathy and a sense of ethical responsibility for others within and across species is not only present in the so-called animal kingdom but also is continuing to evolve in the human species. Primatologist Frans de Waal's pioneering work on the evolution of morality from, and within, our animal ancestors to humans makes evident that animals are empathic and have a sense of responsibility for others, which can be seen as a proto-ethical, if not also an ethical, response. Studies of rats and monkeys indicate that they would rather go without food themselves than witness pain inflicted on others (de Waal 2006, 28–29). Sharing and grooming behavior in animals also indicates a sense of gratitude and reciprocity that could be interpreted as proto-ethical behaviors (de Waal 2006). Following Darwin, de Waal argues that our moral sense or conscience evolved from animal sociability. Furthermore, he maintains that any animal that develops a certain level of intellectual ability will develop moral sensibility (2009, 8). De Waal concludes that we can learn from nature and from animals about empathy and sharing, lessons that can only help us cooperate in our increasingly globalized world (2009). His work suggests not only that empathy evolves within species but also that empathy evolves between or across species. In this case the biosphere is evolving to be more empathetic. Certainly, humans are becoming more empathic toward other animals.

In his survey of intellectual history and contemporary developments in fields across the natural and social sciences, Jeremy Rifkin argues that empathy is evolving among human beings not only for other human beings but also for other species (2009).[12] Rifkin maintains that the convergence of Internet and communication technologies that unite various peoples across the globe with renewable and sustainable sources of energy gives rise to what he calls a "third industrial revolution" based on these technological advances in a synergistic relation to advances in empathic abilities amongst human beings (and perhaps between and among other species) "that could extend empathic sensibility to the biosphere itself and all of life on Earth" (2010). Like de Waal, he believes that we can learn from our animal kin about empathy. Moreover, he argues that the realization that empathy exists in other life on Earth "can't help but change the way we perceive our fellow creatures as well as strengthen our sense of responsibility to steward the Earth we cohabit" (2009, 81). Rifkin provocatively suggests that if we can "harness" this empathy toward others and the earth, we might be able to create a "biosphere consciousness" that is more attuned to our earthly home and our fellow earthlings.[13]

Certainly, increasing concern for animals among human beings, especially in Europe and the United States, signals shifting attitudes toward not only our animal companions but also other animals with whom we share the planet. In the last few decades our attitudes toward animals have changed dramatically. Now, in Europe and the U.S., more people live with companion animals than do not; and many consider these animals part of their families.[14] Laws protecting animals and promoting animal rights are changing to reflect these changing attitudes. If this trend continues, within the next few decades we may live in a radically different world where animals are extended empathy, compassion, and rights never before imagined appropriate for their kind. Indeed, with technological advances in the production of proteins for human consumption, we may see the end of factory farming. And with continued recognition of the importance that companion animals play in our lives, and their positive impact on mental, emotional, and physical health—that is to say our dependence on them—we may see mixed-species households and families as the norm and no longer the stuff of science fiction. Indeed this fundamental change in our relationship to other animals, both particular animal species and animals in general, may be the most significant development of our era.

As we have noted, biologist Edward O. Wilson proposes that human beings have evolved through *biophilia*, which is to say through love of life and love of other living creatures. He argues that the biosphere is a dynamic system in which all parts are interrelated. Extinguish one microorganism and you cannot predict the consequences as they ripple through the ecosystem. Human beings are the result of the great biodiversity of earth. "Biodiversity," says Wilson, "is the frontier of the future" (1993, 39). He identifies biophilia with a "spiritual craving" inherent in our genes that cannot be satisfied through the colonization of space because other planets are not only inhospitable to life but also too far away. There is more life, organization, and complexity in a handful of the earth's soil than on the surface of all the other known planets combined, which is why he concludes, "The true frontier for humanity is life on earth" (1993, 39). We have evolved to love living beings and to be fascinated by other species. But, given rapidly diminishing biodiversity, there is an urgent need for "an environmental ethic based on it" (Wilson 1993, 40). Affirming our dependence on the biodiversity of the biosphere may be a step in that direction, especially if we embrace biophilia as interspecies love. Given that the Greek *philia* is associated with mental love or friendship, however, we need to add the embodied sensual dimension of *eros* in order to imagine an embodied environmental ethic based on our radical relationality not only with other species but also with the earth that supports us all.

Interspecies love may be evolving for the sake of the biosphere. Indeed, the biodiversity upon which our biosphere depends may require interspecies cooperation and interspecies love. Given what human beings have done to destroy ourselves and to destroy the habitats of various species and slaughter others, human attitudes toward our earthly companions need to evolve if we are to learn to share the planet.[15] Our changing attitudes toward other animals signal a new era of interspecies relationships. Certainly the dramatic shift among many people in developed countries to consider companion animals as family, and to love and mourn them, is evidence of the evolution of eros. We could say that the life force is put into the service of interspecies cohabitation. And interspecies cohabitation becomes the ground for ethical responsibility to earth and its inhabitants. Eros is the groundless ground of interspecies ethics and the life force of earth ethics.

In *Interspecies Ethics* Cynthia Willett develops a connection between ethics and eros manifest in her earlier work. Here, focusing on relationships between species and the evolution of ethics from play, she argues that ethics is thoroughly social and develops from play and laughter as ways of facilitating social relations, which are essential to all social animals, including human beings.[16] She argues that the "principleless principle" of ethics is not found in philosophical logos, but rather in playful encounters through eros as a biosocial drive that facilitates bonds between individuals and between species and creates a sense of belonging and home.[17] Willett describes *Eros* as a drive toward home, but not the sentimental notion of home in popular culture or nationalist movements.[18] Contrary to this sentimental notion of home, *ethos* as habitat or home has everything to do with eros, or love as sensuality, which is the social bonding agent that brings creatures together and gives them a sense of belonging (cf. Willett 2014). This drive toward home is neither an individual enterprise nor self-contained within one body. Rather, its means and ends are sociality itself and the bonds that make not just surviving but also thriving possible. Sociality and belonging are tied to earthly cohabitation and our shared, yet singular, bond to the earth.

This sense of belonging resonates with what Heidegger suggests is the pregiven, and therefore unknowable, ground of our experience, namely that we are earthbound creatures. Everything we do comes from the earth. As Husserl insists, everything we experience, whether on the earth or off-world, is *as* earthlings. While the earth is the basis for our perception, understanding, knowledge, and sense of mastery, the earth itself is what cannot be perceived, understood, known, or mastered. And although we share this special bond with other creatures, their worlds both as species and as singular beings are unique in ways that prevent us from ever totally perceiving, understanding, knowing, or mastering them. In this earth ethics, an extended Arendtian plurality of worlds makes up the biosphere, which is held together through biophilia, or the love of different living beings. Through this biodiversity and cultural diversity, we share an inherent bond to the earth, and the need to belong to it, as well as to our own world(s) grounded on it. Animals too have their own cultures, which contribute to the cultural diversity and plurality of worlds on earth.[19] Moreover, the diversity of worlds and cultures on earth contribute to our uncanny home, both familiar and strange, but certainly where we belong. While earth may resist and refuse attempts to assimilate it into a notion of *home* as completely known and familiar, it grounds our sense of home as the uncanny mystery of cohabitation on this "pale blue dot," "this island earth," this lovely "blue marble" that we all necessarily make our home, whether we literally call it *home* or merely live by virtue of our connection to it. Even those of us who do call it *home* (or *Heimat*, *maison*, *casa*, or *hjem*, etc.) also live by virtue of belonging to the earth in ways unknown and unknowable to us. For Heidegger, home is a relationship of *belonging* as dwelling with, which means "to cherish and protect, to preserve and care for" (1971a, 349).[20]

To say that we are earthbound creatures is to say that we have a special bond to the earth. We belong to the earth, just as it belongs to us. Rather than ownership, this sense of belonging harkens back to a more archaic sense of the word that conjures eros as longing and companionship. Our life on earth is a longing for home, for a home that we can love, a home that we love enough to preserve and protect.[21] Not, however, in the sense of homeland security or management.[22] Rather, in the sense of letting it come into its own by respecting the ways in which it is not of our making and out of our control. This sense of belonging as *freeing* reverberates throughout the arc that we have been tracing from Kant to Heidegger. Here this belonging is figured as eros, a connection to those around us through which we come into our own and the responsibility to let them be so that they can come into their own. Insofar as it is essentially relational, freeing is born out of eros rather than recognition or reason.[23] The interconnectedness of all life on earth, combined with the affective bonds of social animals, make our relationality primary. It preexists recognition and reason; it is the prior connection and connectivity that makes them possible. More generally (and less anthropocentrically), we could say that *home* is, at the same time, given and yet the result of relationships with alterity, by virtue of which we not only live but also thrive and through which we both belong and come to belong. Home as belonging, then, could be seen as both the means and the ends of an ethos of earth.[24]

Ethos as habitat or home brings with it a sense of belonging to an ecosystem or community.[25] This sense of belonging is not a familiarity that can be taken for granted, especially when we consider earth as home. For, as every creature "knows," the earth is populated with strange others and foreign landscapes that can be welcoming or threatening, and everything in between. For human beings the earth as home is fore-given and must be interpreted and reinterpreted, even as it is also a prerequisite for meaning. Willett describes "biocultures of meaning" based on social bonds between companions, places, memories, histories, and interspecies relationships, what Derrida might call the *miracle* of social bonds that build bridges between living beings (Willett 2014, 23; cf. Naas 2014, 51–52). Meaning both requires social bonds and emerges through social bonds, which are tied to particular spaces or places and times or histories. The relationality of social bonds, including bonds to places and histories, makes meaning possible, even while meaning emerges through relationships.

The dynamic of meaning as both constituted by, and constituting, our relationships are akin to what I call the *witnessing* structure of response-ability, the structure of address and response (see Oliver 2001). Living creatures are responsive, and an earth ethics promotes our responsibility to open up, rather than close off, the response ability of others, their ability to respond. Even in the face of our lack of understanding, the impossibility of mastery, and inherent unpredictability, we have a responsibility to act in ways that open up the possibility of response from our fellow earthlings and from the earth itself. Obviously this abstract "principleless principle" or "groundless ground" also opens onto the tension between ethics and politics discussed throughout this book. Ethics requires that we open up response and response ability in the face of our ignorance—for, if we knew with certainty, it would no longer be ethics but social or even natural science. While politics requires that we negotiate relationships within our living space in order to survive and thrive, which always necessarily means killing or excluding some others (e.g., deadly bacteria, fungus, and viruses). Again, we might say that an ethical politics is one in which ethics juts through political policy and forces us to continually and vigilantly reassess and reinterpret our responsibility toward others, even if—perhaps especially if—those others are threatening.

Expanding on the ethics of response, Willett argues for an ethics of "call and response" in order to recognize the vocal communication and expression of animality in both nonhuman animals and human animals.[26] Importantly, call and response also can refer to interspecies communication, which expands the notion of witnessing to nonhuman animals and perhaps even beyond if we take a broad enough view of response. In *Animal Lessons*, for example, I expand the notion of witnessing to include nonhuman animals. The basis for ethical relations has moved beyond reason or recognition and toward witnessing to response ability itself, that is to say witnessing to the ability to respond, which is not just the domain of humankind, but of the entire "animal kingdom." In this way, witnessing ethics as response ethics can take us beyond human centrism and toward consideration of the ways in which all creatures of earth and earth itself respond. Within response ethics, political and moral subjects are constituted not by their sovereignty and mastery but rather by address and response. Extending my analysis of witnessing, address and response (broadly conceived) are the basis of earth ethics grounded on cohabitation and interdependence. And the responsibility to engender response, or facilitate the ability to respond, in others and the environment, is the primary obligation of earth ethics. This earthly ethos is the result of pathos beyond rationality or recognition because it is based in our embodied relationality, which is bound to other living beings not only through shared places and histories but also through the larger biosphere and ecosystems that sustain us, and ultimately through our singular bond to the earth.

"HOME IS A MEMORY WORTH FIGHTING FOR . . ."

If ethos is grounded in eros rather than in logos, then what does it mean to talk of an *ethos of the earth*? If ethos as eros is a drive toward home, do we have a drive toward earth as our home? As Heidegger suggests, is a sense of home only possible from a distance, which means that home is always infused with melancholy and nostalgia? If popular filmic images of the destruction of earth and the melancholy and nostalgic endeavors to regain it are any indication, then, within our cultural imaginary, our ambivalent relation to our earthly home is manifest through its destruction and reconstruction over and over again. A detour through one such recent narrative, *Oblivion*, may be instructive.

The film *Oblivion* is another recent filmic journey home to earth. Director Joseph Kosinski's postapocalyptic vision of a future, where "we used the nukes" and the earth is turned into a nearly uninhabitable desert, combines anxieties over the nuclear threat so powerfully felt in the cold war and concerns over environmental devastation so powerfully felt today. Filled with nostalgia and longing for earth, the protagonist, flyboy Tech 49, Jack Harper (Tom Cruise), secretly recreates paradise on earth when he builds a cabin in an oasis of evergreen forest nestled on the shore of a beautiful lake. "Earth," as the tagline of the movie tells us, "is a memory worth fighting for."

Through a series of plot twists, Harper discovers that it was not "the nukes" that ruined the planet, but rather the alien invasion from outer space. Once this revelation is made, it is easy for Harper to fight against this enemy from outside in order to protect his home, earth. Within the narrative of the film, Harper goes from blaming his own species to blaming nonhuman aliens for the environmental disaster that the earth has become; of course, it is always easier to fight an enemy that we consider foreign, even monstrous, than it is to confront the "enemy" within. For, nothing unifies a people faster than a common enemy. Is it possible, then, to see man-made climate change and environmental pollution as the common "enemy"? Could this be the "saving power" of global technology?

Oblivion is a story of nostalgia and longing to return to earth and to return vitality to earth. Harper is a Robinson Crusoe–type figure that must start again from scratch in his hidden forest oasis. Perhaps more than either *All Is Lost* or *Gravity*, *Oblivion* shows our ambivalent relationship to earth and to our own position on it. Like the hero of *All Is Lost*, Tech 49 dies in the end. Unlike "our man," however, Harper's Tech 49 does not die in vain, apologizing for being a failure, but rather he sacrifices himself to defeat the aliens and return humanity to earth. He sacrifices himself for his wife and unborn daughter (a long story), whom he drops off at his cabin oasis before heading on his kamikaze mission. Like "our man," Harper is a maverick self-reliant character that puts up a manly fight against the element and the aliens. He uses his wits to outsmart the technologically dependent disembodied aliens. Unlike "our man," however, Harper turns out to be a clone and not an individual at all. Indeed, there are countless replicas of Harper, none of which are the "original." Furthermore, it is Harper's connections to "his" latent memories (which, of course, are not his own) of people and places he loves that sustain him and inspire him to fight for those memories, the memory of earth. It doesn't matter that he is not a unique individual or that the memories are not his own; they are real nonetheless. And the connections to people and places, eros and longing, cannot be repressed by technology. In the end, he sacrifices himself for love, a seemingly eternal love that is reborn with each clone. In the final scene, Tech 52 Harper makes it to the forest refuge and finds his wife and daughter waiting there for him. Whether he is a copy or an original is irrelevant to the emotional bond he has to his wife, his forest, and ultimately to the earth. So, while he may appear to be the maverick individual familiar to us from both our popular and intellectual history, there is a twist in *Oblivion* that turns and returns "our clone" to his earthly home. Ultimately, it is his embodied connection to the earth that allows him to defeat the disembodied aliens and reestablish his home on earth. Home, as we learn in *Oblivion*, in the sense of belonging, is lived as embodied interrelational beings driven by eros in a drive toward home. And yet, even when it has not been rendered a desert by aliens from outer space, our earthly home is *unheimlich* or uncanny, and we have an ambivalent relationship to it, as evidenced by so many fantasies of its destruction.

THE MISSION TO EARTH

Encounters with the unfamiliar, or the familiar become strange, can conjure the uncanny sensation associated with thinking about earth as home.[27] The earth's diversity confronts us nearly at every turn. Because of earth's diversity, and the uncertainty of our place in relation to it, not to mention our limited perspective on it, contemplating earth can be overwhelming. While in one sense the earth is familiar to us, as familiar as anything can be, it is also always unfamiliar insofar as we encounter its exhilarating strangeness whenever and wherever we look for it and most especially when we are not looking. Maybe this is why, on those extraordinary missions to the moon, it is the Earth that came into view. While we were "not looking," we saw the Earth anew, astonishing as seen from space for the first time.

From this distance, reactions to seeing Earth were ambivalent, already steeped in nostalgia for returning home and melancholy loneliness faced with the vastness of the universe. Coming in the midst of fears of nuclear destruction, these images of the "whole" Earth conjured their counterpart in the iconic mushroom cloud and filmic representations of the destruction of earth, first through dreaded nuclear war and eventually through its evil twin, environmental catastrophe. It is as if, in order to make earth meaningful, we must imagine ourselves without it. Indeed, insofar as we rely on photographs of Earth, which are themselves two-dimensional representations that "capture" only partial views of the whole, our relationship to the whole earth is always necessarily limited and mediated. These representations of earth signal ambivalence already provoked by our limited perspective itself. These images are uncanny not only because they are taken from the impossible position of an astronaut floating in space but also because they are taken from the inherently limited human perspective, mediated by technologies that flatten the earth and reproduce it endlessly on glossy paper and computer screens. In this way the photographs of Earth from space already suggest the annihilation of earth and incite our ambivalence toward our own limited position on earth. Indeed, there is still so much that we don't know about the Earth and our solar system. The photographs of Earth from space remind us of this fact. They continue to conjure the uncanny sensation of both our expansive possibilities and our human, all too human limitations. On the one hand, we are by far the most technologically advanced species on the planet. On the other, we still don't know how to stem the environmental that we have caused through our very technological prowess.

But, insofar as the environmental crisis has the potential to bring us together as a species with the common goal of saving the earth, this may be the "saving power" inherent in the technological worldview. Perhaps technology could engender some less romantic and less abstract version of the unification of humankind imagined during the Apollo missions. Certainly, technology is bringing distant lands closer together. Internet technology gives many people almost instant access to different cultures and countries. Transportation technology allows us to travel *almost* anywhere on earth, at speeds unimaginable a century ago. Keeping in mind the dangers of the "McDonaldsification" of the entire globe threatened by globalization, which reduces the entire world and earth to markets, can we also imagine the positive possibilities of globalization? In other words, can we see globalization in all its ambivalent glory? Heeding the warnings of Kant, Arendt, Heidegger, and Derrida, can we imagine the positive potential of global cosmopolitanism put into the service of the earth? At this moment, with globalization in full swing, it is important to conjure the uncanny moment that has become emblematic of the beginnings of both globalization and environmentalism, the moment that the Apollo missions to the moon transmitted photographs of Earth back to their terrestrial home. This moment is emblematic of both the dangers and the saving power of technology and of our singular bond to the earth.

Perhaps the "saving power" inherent in global technology has the potential to engender a cosmopolitanism that takes us beyond international laws and tribunals and toward cooperation based on shared responsibility for the planet by giving us a common purpose, "saving" the earth. Already, albeit with limited success, there have been international efforts to curb climate change and control man-made pollution.[28] Since the images of earth were broadcasted from space, and the environmental movement was born, many people across the planet have been working in various ways, individually and in groups, within national borders and beyond, to limit man-made pollution. Even if those efforts have yielded only minimal results, the environmental crisis is on the radar now in ways that it wasn't decades ago, thanks to growing concern about climate change and mounting evidence about the threat that it poses not only to human life but also to other life forms and to the general health of the planet. If the saving power of the environmental crisis is that it has the potential to unite peoples across the globe and bring them together with a common cause, it also has the potential to unite us with other species beyond our own. It may even have the potential to change the way we conceive of our relationship to other species and to the earth, if we begin to see ourselves as earthlings first and foremost. We see that we are earthlings who share this planet, our only home, with countless other earthlings who, like us, are profoundly dependent upon sharing our singular earth. We realize that, with all other earthlings, we share a singular bond to our home planet. Possibly this is why the images from the Apollo missions still conjure our uncanny position in the universe as earthbound beings reaching for the stars.

Rather than uniting all humankind through fantasies of planetary destruction or the end of the world, or through global technologies of mastery and global markets that render everything and every one of us fungible, we can imagine an ethics of earth that gives rise to connectedness beyond the autonomous moral subject, beyond humanism, and beyond recognition. This is what it would mean to belong to the earth as the home we share with all living creatures. This ethos of the earth can provide the grounds for a nontotalizing, nonhomogenizing earth ethics, if we can imagine a dynamic ethics based on the response ability of biosociality and biodiversity rather than on universal moral principles that may close down the possibility of response. Earth ethics opens rather than closes the possibility of response and response-ability. In this way, earth ethics operates like Heidegger's *poi-*

esis or Derrida's *poetic as if* in order to open onto the alterity of earth rather than use it up in one totalizing worldview. Earth ethics can become a counterbalance to the totalizing tendency of globalization. Rather than the planetary imperialism or planetary calculation that Heidegger warns uproots us from earth, returning the global to the earth reminds us that it is the earth on which we live. We do not live on the globe, but rather on the earth. The earth is a counterbalance to the globe. Perhaps we can envisage the photographs of earth from space, lovely and alone, as the catalyst for *amor terra* rather than global management or global technology. *Amor mundi* based on *amor terra*—love of world based on love of earth—may transform the earth from the meaningless and solitary desert island feared by Arendt, Heidegger, and, in his own way, Derrida, into our uncanny home, as unsettling and mysterious as it is necessary for life.

Retracing our steps one last time, we started with the realization that we all survive only by virtue of sharing the limited surface of the earth, which obligates us to form political alliances and economies. Given that we do not, and cannot, choose with whom to cohabit the earth, however, our freedom to choose political alliances presupposes the unchosen character of our coexistence. Furthermore, biodiversity and cultural diversity allow each individual and each people or species to thrive. And the plurality of worlds, peoples, and species fill the world with promise and surprise, which not only thwart our attempts at totalization and mastery but, moreover, open up the possibility of freedom and the wonder of relationality. Each singular being is completely unique—at the extreme, a world unto itself—yet shares the miracle of cohabitation. It is not just that we share physical space, or proximity, on the surface of the earth, but, more significantly, we share a special bond to the earth as our only home, whether home or habitat. This singular bond of all living beings to the earth and to other earthlings directly, and indirectly, obligates us to the sustaining possibility by virtue of which we not only exist and survive but also live and thrive. Through our relationships with others, human and nonhuman, organic and inorganic, our earthly cohabitation is imbued with meaning.

Our dependence on those others can be the source of a notion of the globe grounded on an earthly ethos as ethics of response ability, or it can be reduced to the totalizing fantasy of control and mastery, whether technological globalization, global marketing, or even global management for the sake of the environment. The fact that we exist and become human through our fundamental interconnectedness and social bonds can lead to a sense of gratitude toward others as thankfulness (in the sense of thinking as thanking) and the freedom to let each come into its own by responding through listening, thereby engendering the possibility of response. Or this primordial indebtedness can be exploited in global markets that reduce all relationships to debt, calculation, quantification, and exchange values. Tensions between earthly forces, between earth and world, between world and world, between different peoples or species can be productive and generative of change and creativity. Or they can become the rationale for violent conflict over property and control. If the limited surface of the earth only grounds property rights, then, any hope of returning to earth is replaced by global ownership. On the other hand, if we see *belonging* not in terms of property but rather in terms of the longing and companionship of its archaic meanings, then our belonging to the earth is born from our singular bond to earth. Belonging as longing. Eros is the drive toward home that returns the globe back to the earth.

Caring for home, however, necessitates avowing our ambivalence toward our earthly habitat in order to embrace, rather than annihilate, the uncanny queasiness of belonging to this planet with multitudes of others, most of whom we have never encountered. Affirming our singular bond to the earth takes us beyond the earth's limited surface, or the plurality of cohabitation, and toward the groundless ground of our uncanny earthly home. As we have seen, the earth is much more than its surface or its peoples (or animals). Indeed, the Earth's surface is much more than it seems, teaming with life such that one handful contains more than all the other known planets combined. The meaning of *earth* is a complex of relationships, most basically, between its biosphere and the other spheres that make it a living planet. In this sense, we have only begun to scratch the surface of what science may discover about the dynamics of earth. Whatever we discover, earth is our home, and a home that we share with multitudes of other earthly creatures. All earthly creatures have a singular bond to the Earth that makes it more than one planet among others. It is this singular bond of each and every one combined with the fact that we share this bond that ground earth ethics. The singularity of each in its unique relationship to its environment contributes to our shared earth. Rather than detract from what we share, namely this singular bond to the earth, the plurality of worlds constitutes the meaning of earth as a network of relationships. An ethics of earth is grounded in the affirmation of social and biodiversity that makes the earth a living planet. Earth ethics emerges from the tension between the absolutely unique place of each one and the collectively shared bond to the earth, both of which necessarily constitute the life of the planet. Earth's biosphere, which cannot be separated from the earth itself, is a dynamic of individuals and communities, species and interspecies symbioses. And all life is dependent upon nonorganic elements that are also terraforming. The earth is this complex of relationships. And, for us, from our uniquely human standpoint, the earth has both the prereflective meaning of our embodied connections and the more poetic meanings that result from meditations on our earthly dwelling. The realization that earth is our one and only home fills us with ambivalent reactions: on the one hand, we feel confident that we hold a unique and special place in the universe and, on the other, we feel insecure and insignificant relative to the vastness of the cosmos.

Returning to the vantage point from which we began, that is to say, the impossible vantage point of the Apollo astronauts, our earthly home appears fragile, vulnerable, and alone. And yet it also appears whole, beautiful, and unique. The ambivalence of this unsustainable view from space is emblematic of the ambivalence of the human perspective on earth and world more generally, a perspective that is always partial and limited and yet longs to be complete and boundless. On the one hand, the technological achievement of launching ourselves into space and leaving earth's atmosphere leads to feelings of mastery and control. On the other hand, seeing the Earth as a "tiny pea" from space leads to feelings of insignificance and alienation. As we have seen, the ambivalence of the discourse surrounding the Apollo missions is evidenced by the seeming contradiction between slogans of "America First" and "uniting all mankind," both of which were used in connection with the missions.[29] In addition, the photographs of Earth sparked concern for both global technological access and a global environmental movement. Whatever ideological work the rhetoric of "uniting mankind" was doing for the NASA missions, the moments when the "people of earth" (at least those living in technologically advanced cultures) first saw images of Earth floating in space, however, were filled with a sense of awe and uncanny closeness. Indeed, those images still send shivers up the spine, especially when combined with the now iconic voice of Carl Sagan describing that "pale blue dot" as home.

GRAVITY

The recent film *Gravity* revels in stunning images of earth from space that can't help but remind the viewer of the now pervasive *Earthrise* and *Blue Marble* photographs shot on the Apollo missions to the moon. Indeed, this film is a testament to the continued importance of those photographs in our cultural imaginary. Like the Apollo missions to the moon, even though it takes place in space, *Gravity* redirects our attention to the earth. Within the narrative of the film, medical engineer Dr. Ryan Stone (Sandra Bullock) is on her first shuttle mission with veteran astronaut Matt Kowalsky (George Clooney), trying to repair the Hubble Space Telescope, when debris from an exploded satellite destroys their shuttle and sends them hurdling into space. Kowalsky reaches Stone, and, tethered to each other, they try to reach the International Space Station using only a thruster pack. Kowalsky sacrifices himself to allow Stone to reach the space station. The rest of the film shows Stone encountering one disaster after another as she resolves to do what it takes to return to Earth, but only after a pep talk from Kowalsky's ghost (or her hallucination due to oxygen depletion). Like "our man" from *All Is Lost*, Dr. Ryan Stone is emotionally detached. We learn that her emptiness and lack of even the will to live is the result of the accidental death of her young daughter years before. Unlike "our man," she is transformed and redeemed by her near-death experience and her struggle to survive.

If "our man" represents the bygone days of the liberal subject, self-contained and independent, Ryan Stone is so connected that she has lost herself in her melancholy relation to her dead daughter. Like "our man," she is tethered to her technology to keep from floating away. But, it is her transformation that saves her and not the technology alone. Indeed, it is the visitation by Kowalsky that convinces her to live and fight to survive. He convinces her not to give up. And if "our man" is going it alone with signs of globalization passing him by, Stone is a citizen of the world insofar as she uses technology from the U.S., Russia, and China in order to go home. Technology operates as a kind of universal translator that enables her to move from ship to ship, using various manuals in languages that she does not understand. Even when she doesn't understand the Inuit fisherman whom she hears cooing to his baby over her radio, she is comforted by his presence; and the barking of his dog and his voice become a lullaby of sorts as she prepares for her final sleep. And, yet, it is not technology, but rather the ghost/hallucination of Kowalsky and the memory of her daughter that enable her survival. There is no hand of God to reach down and save her, only a spiritual connection to those around her, transferred in the end to the earth itself. Ultimately, it is earth's gravity itself that pulls her back into its embrace.

In a tense ninety-one minutes, director Alfonso Cuarón creates a frenzied attempt to return to earth that ends with Stone gasping for air and clutching the sand, desperately grasping at earth. She emerges from the water like the first amphibious life crawling out of the primordial swamp, alone and yet reunited with earth. Stone's return to earth is symbolic of our need to return to earth. Like that of any other animal, her body is bound to the earth, which the last scene captures in all its gravity. Unlike "our man," who is shown tiny and insignificant as he sinks deeper into the sea, "our woman" towers as she ascends from the water, an Amazon filling the big screen. Together, *All Is Lost* and *Gravity* show us the two sides of our ambivalent relationship to Earth, namely that we appear huge and important, the center of the universe, and that we are puny and insignificant, no more than minuscule specks of space dust. And while *Gravity* shows how small we can be when floating in space, disconnected from Earth, it also shows how monumental the return home can be once we learn to appreciate our connection to earth.

One message of the film is that in spite of our losses, which are considerable, when the world is gone, we remain connected to others and to the earth. And if we can embrace that connection, if we can love it enough to take responsibility for it, then there is hope that we might reach home again. In a global world, linked by satellites circling the earth, our technological and spiritual connection to others—even to barking dogs—gives meaning to life, which takes us beyond mere survival. As the film warns with its opening voice-over, "no one can survive in space." And we need that reminder. The warning that no one can survive in space, and Stone's symbolic return to the earth, powerfully elicit the most exhilarating moments of human space exploration, namely the return home. Surviving the risky voyage into space is dangerous, but making it back to earth, not only without being incinerated but also without losing hope and a reason to live is even more perilous. As *Gravity* shows us, it is not just surviving that counts. In fact, its shows us that survival is not possible without love.

Stone's transformation is possible only because, thanks to Kowalsky, she opens herself up to love and longing, both for people around her (or her memories of them) and for earth. It is as if her lost colleague infuses her with his love of life and his love of earth, even with his last living words, reveling in the beauty of the sunrise over the Ganges River. Whatever the problems with *Gravity*'s "new age" spirituality, it makes evident that, unlike "our man," "our woman" does not go it alone. Even when she is alone, she is not alone. No woman is an island. And, "our woman" is not going to be mistaken for the maverick self-contained individual asserting his mastery over both earth and world. Rather, she is saved by listening and responding, perhaps even poetically remaking the world, as if she could still hear Kowalsky's voice encouraging her. She decides to love the world enough to take responsibility for the memories of those she loves, even when her world is gone. And, like *Oblivion*'s Jack Harper, she decides that earth is a memory worth fighting for.

Moving from one mechanical womb to another, space shuttle to space capsule, shown floating in a fetal position, Stone is born and reborn into the chaos of space. But it is not until she is reborn on earth that she is truly saved. Unlike "our man," rising from the waters that threaten to swallow her, she crawls to shore, her body too huge and heavy from the effects of gravity to support itself. Determined, she stands triumphantly at the end, breathing the air and uprighton the solid ground of earth. Like one of earth's elements, Stone becomes extraordinary as she takes her first steps back on earth. Even her name, *stone*, evokes the earth and its elements. Stone is necessarily and thoroughly an earthbound being. She is of the earth and must return to it. As the film reminds us, she cannot survive in space. She belongs to earth, her one and only home.

Stone's harrowing journey back to earth recalls so many times we have seen space modules returning to earth, burning through earth's atmosphere, pulled back down by gravity. With its stunning computer generated sets, *Gravity* recaptures the dizzying view of Earth from space on the first missions to the moon. For ninety minutes the viewer is suspended in the uncanny space outside the Earth's atmosphere. With its haunting silent space walks, and astronauts stranded in space, the film conjures the uncanny sensation of floating outside of our earthly home evoked by the Apollo images. *Earthrise* and *Blue Marble* haunt *Gravity* from beginning to end. The response to this captivating film reminds us of the relevance of the Apollo photographs of earth from space. For it is not the moon that continues to captivate us decades later, but rather the images of Earth, taken almost by accident, on those first moon missions.

NOTES

1. THE BIG PICTURE

1. See Benjamin Lazier (2011, 606). I am fortunate to have found Lazier's article "Earthrise" while working on this project (2011). This chapter is indebted to his work there. I am also grateful for conversations with Jennifer Fay, which helped me immensely in formulating this project.
2. See Lazier 2011, 620; and Cosgrove 1994, 272.
3. See Sagan 1994; and Nicks, *This Island Earth*, 1970.
4. See Lazier 2011, 606. Lazier makes this point and coins the phrase "the globalization of the world picture."
5. Quoted in Poole 2008, 2 (my emphasis).
6. See Derrida 2005c.
7. See Arendt 1966.
8. Kant 1996c [1795], 338.
9. See Benhabib 2006, 17–18.
10. For a discussion of the complexities of Arendt's thoughts on International law, see Benhabib 2006.
11. Biologist Edward O. Wilson says, "There is no question in my mind that the most harmful part of ongoing environmental despoliation is the loss of biodiversity" (Wilson 1993, 35).
12. In his *Anthropology* Kant claims that ultimately we cannot understand ourselves as rational beings until we meet rational beings from another species. And since for Kant no other species on Earth is rational, he imagines meeting rational extraterrestrials (2006, 225, 7:321). Kant also imagines that all planets are inhabited or will someday be inhabited (2012c, 297, 1:355).

For a discussion of all Kant's aliens, including women, see Clark 2001.
13. Along with many more, including *This Is Not a Test* (1962), *Atomic Rulers of the World* (1964), *Fail Safe* (1964), *The End of August at the Hotel Ozone* (1967).
14. Lazier makes this argument; see 2011, 619.
15. See Cosgrove 1994. For a discussion of the rhetoric of the missions in terms of gender, see also Garb 1985.
16. Denis Cosgrove describes the way in which this panhuman rhetoric aligns Christian universalism and the American vision of global harmony imagined because imperialism can be taken into space where there is enough to go around (1994, 281). See also Poole 2008.
17. See Cosgrove 1994, 287.
18. Quoted ibid., 286.
19. Quoted ibid., 282.
20. Lovell quoted in *Time* 1969, 12.
21. Anders says, "The Earth looked so tiny in the heavens that there were times during the Apollo 8 mission when I had trouble finding it. . . . I think that all of us subconsciously think that the Earth is flat or at least almost infinite. Let me assure you that, rather than a massive giant, it should be thought of as the fragile Christmas-tree ball which we should handle with care" (quoted in Cosgrove 1994, 284).
22. Nicks 1970, vi (my emphasis). Upon seeing the photographs of earth from space, news anchor Walter Cronkite described the Earth as "*floating* in space" (Poole 2008, 146 [my emphasis]).
23. George Low, in Nicks 1970, iv (my emphasis).
24. Quoted in Poole 2008, 2.
25. For a discussion of Derrida's analysis of islands in relation to Kristeva's notion of abjection, see Negrón 2011.
26. See Garb 1985, 21.
27. Quoted in Poole 2008, 20, see also 135.
28. It is noteworthy that while some of the geographers and historians who have discussed the Apollo photos quote Edmund Husserl on pre-Copernican Earth or point out that Blue Marble only shows Africa and Asia; none linger on the fact that these photographs are actually not of the *Whole* Earth.
29. For helpful discussion of this essay, see Himanka 2000 and 2005.
30. See Sallis 1998, 206.
31. Claire Colebrook also mentions Lacan's mirror stage in relation to the whole earth (2012, 31).
32. Cf. ibid., 32.
33. For one psychoanalytic approach to analyzing the fantasy of wholeness inspired by the photographs of Earth, see Bishop 1986.
34. Lazier gives an illuminating account of reactions to the photographs of Earth from space, which includes some discussion of logos and icons based on the photographs (2011).

35. For a discussion of the dangers of globalism in terms of world picture, see also Nancy 2007.
36. Arendt and Jaspers, 1993, 363.
37. See Lazier 2011, 603.
38. For a discussion of hospitality in Derrida, see Westmoreland 2008.
39. Derrida 2003a, 98–99.
40. According to United Nations economist Gao Shuangquan, "The difference of income per capita between the richest country and poorest country has enlarged from 30 times in 1960 to the current 70 times. And over 80% of the capital are flowing among US, Western European and East Asian countries." Shangquan 2000, 4). For a more detailed discussion of how globalization has created a widening gap between the global North and the global South, see Manfred Steger Globalization (2003).
41. For a discussion of Derrida's preference for worldwide over global, see Li 2007.
42. Cf. Heidegger 2012a, 48–49.
43. Heidegger 1971d [1951], 217; see also Heidegger 1971a [1951]; cf. Mitchell 2011, 12–13.
44. Mitchell 2005, 202; see also Heidegger 1954 "Overcoming Metaphysics."
45. "Evening Conversation in a Prison of War Camp in Russia" (1945), quoted in Mitchell 2005, 20n, 217.
46. In his thought-provoking essay on Heidegger and terrorism, Andrew Mitchell describes this desertification: "Devastation (*Verwüstung*) is the process by which the world becomes like a desert (*Wüste*), a sandy expanse that seemingly extends without end, without landmarks or direction, and is devoid of all life.... The lifeless desert is the being-less unworld from which being has withdrawn.... Yet this unworld is not the opposite of world; it remains a world, but a world made desert" (Mitchell 2005, 206).
47. Heidegger quoted in Mitchell 2005, 206.
48. Cf. Arendt 1954, 273. See chapter three where I discuss Arendt's criticisms of Einstein's "observer who is poised freely in space."
49. For a different approach to this question, see Schroeder 2004.
50. Arendt 1954, 196. Thanks to Anne O'Byrne for bringing this passage to my attention.

2. THE EARTH'S INHOSPITABLE HOSPITALITY

1. For discussions of Kant's notion of hospitality, particularly in relation to Derrida's, see Brown 2010 and Nursoo 2007.
2. Peter Fenves's *Late Kant: Towards Another Law of the Earth* is a fascinating study of Kant's changing notion of earth (Fenves 2003). Ultimately, Fenves argues that Kant concludes that the earth does not belong to human beings, but rather we are preparing it for its true trans- or posthuman owners. Otto Reinhardt's essay "Kant's Thoughts on the Ageing of Earth" is also useful in thinking about Kant's

theory of earth (Reinhardt 1982). See also Reinhardt's discussion of Kant's analysis of earthquakes and volcanoes (Reinhardt and Oldroyd 1983).

3. For a discussion of Kant on the human species, see Cohen 2006. For an interesting analysis of the relationship between human nature and Kant's moral theory, see Edwards 2000; see also Kain 2009. For a discussion of the perfection of the human species, see Zammito 2008.

4. In "Conjectures on the Beginning of Human History," Kant says, "nature has endowed us with two distinct abilities for two distinct purposes, namely that of man as an animal species and that of man as a moral species" (Kant 2003, 228n).

5. See ibid., 224–225.

6. For helpful discussions of Kant's notion of cosmopolitanism, see Hedrick 2008; Linklater 1998; Louden 2008; Kleingeld 1998, 2003; Wilson 2006, 2011; and Wolin 2010. Wolin extends Kant's analysis to talk about contemporary issues of globalization. Hedrick addresses the limitations of Kant's notion of cosmopolitanism in terms of multiculturalism and race. Wilson puts Kant's notion of cosmopolitanism in the historical context of his pedagogy. Kleingeld shows how Kant argues for both patriotism and cosmopolitanism and world citizenship. Linklater uses Kant to argue for a dialogic approach to citizenship. Louden argues that Kant proposes a cosmopolitan notion of human nature.

7. Kant claims "that one animal species was intended to have reason, and that, as a class of rational beings who are *mortal individuals but immortal as a species*, it was still meant to develop capacities completely" (Kant 2007b [1784], 44).

8. In his careful analysis of Kant's property law in "The Unity of All Places on the Face of the Earth," Jeffrey Edwards argues that Kant's turn to a spatial justification for private property based on original common possession of the earth's surface must be consistent with the innate principles of freedom, equality and independence that Kant sets out as fundamental to the doctrine of right (Edwards 2011, 257).

9. Edwards cites this passage from Kant's loose leaf: "this possession must also be regarded as collectively universal, i.e., as the common possession of the human species to which corresponds an objectively united will or will that is to be united; for without a principle of distribution (which can only be found in the united will as law) the right of human beings to be anywhere at all would be entirely without effect and would be destroyed by universal conflict" (quoted in Edwards 2011, 244).

10. For analyses of Kant's theories of property and possession, see Byrd and Hruschka 2010; Westphal 1997, 2002. Westphal argues that Kant's theory of property assumes rather than proves the legitimacy of possession (2002). See also Skees 2009.

11. Jeffrey Edwards wrote this in a personal e-mail correspondence. For a discussion of the relationship between human nature and human freedom, see Sturm 2011.

12. Kant says, "Hence, under the general concept of public right we are led to think not only of the right of a state but also of a *right of nations (ius gentium)*. Since the *earth's surface* is not unlimited but closed, the concepts of right of a state and of a right of nations lead inevitably to the idea of a right *for a state of nations (ius gentium)* or *cosmopolitan right (jus cosmopoliticum)*. Kant 1996a, 455, 6:312 (my emphasis).

    He concludes, "The rational idea of a *peaceful,* even if not exactly friendly, thoroughgoing community of all nations on the earth that can come into relations affecting one another is not a philanthropic (ethical) principle but a principle *having to do with rights,* Nature has enclosed them all together within determinate limits (by the spherical shape of the place they live in, a *globus terraqueus*). Kant 1996a, 438, 6:352 (my emphasis); cf. 1996a, 404, 6:250 ff (my emphasis); see also 1996c, 326, 8:355.
13. If equal access and not equal distribution is what is required, then Edwards's argument is not quite as strong, which may very well be the case.
14. See Kant 1996a, 416, 6:265.
15. See Kant 1996a, 420.
16. In answer to the question of whether developing the land is necessary for acquisition, Kant answers "No." See Kant 1996a, 417, 6:265.
17. See Kant 2011 [1760], 7:127.
18. For more examples of animal resistance to humans see Hribal 2011.
19. See Kant 1996a, 419, 6:268.
20. According to Vernadsky, the first was Professor E. Suess in 1875 in *The Origin of the Alps.* Jean-Baptiste Lamarck introduced the term *biosphere* in 1802 in a book entitled *Hydrogéologie.* See Vernadsky 1998, 91.
21. On the usefulness of earthquakes, see Kant 2012a, 360, 1:456.
22. Compare Kant 2006, 427, 7:331.
23. See Kant 2009, 298, 1:355–356.
24. See Kant 2003, 224.
25. For Derrida's discussion of Kant's claim that war is necessary, see Derrida 2011, 272–273.
26. See ibid., 9:165.
27. For interesting examinations of Kant's relation to extraterrestrials, see Dick 1982 and Clark 2001. David Clark's essay "Kant's Aliens" is an engaging and detailed account of Kant's remarks about extraterrestrials in relation to his theory of "man."
28. As Anne O'Byrne reminds us: "the Germanic word *belonging* in its earliest sense means to go along with, to be proper accompaniment to, to be appropriate to. Then the verb *to belong* emerged in English as an intensification of *to long,* thereby incorporating an element of desire and necessary separation. Finally, the archaic adjective *belong* has its roots in equality. The Oxford English Dictionary tell us: '[t]he primary notion was apparently "equally long, corresponding in length," whence "running alongside of, parallel to, going along with, accompanying as a property or attribute"' (O'Byrne 2013).

3. PLURALITY AS THE "LAW OF THE EARTH"

1. For example, see Chapman 2004, 2007, Macauley 1992, 1996; Mortari 1994; Ott 2009; Smith 2011; Szerszynski 2003; and Whiteside 1994.
2. For discussions on Arendt and limits (or lack thereof), see Canovan 1983; Disch 1994; Gottlieb 2003; Wolin 2003.
3. For a helpful analysis of freedom in Arendt, see Kateb 1977.
4. For an account of Arendt's life, see Young-Bruehl 2004.
5. For passing discussions of the notion of belonging in Arendt's writings, see Butler 2011, 2012a.
6. Julia Kristeva emphasizes Arendt's notion of narrative in her study of Arendt (Kristeva 2001, 99).
7. For an engagement with Arendt on human rights that argues for reconceiving human rights, see Burke 2008. See also Birmingham 2006; Benhabib 2007; Brunkhorst 1996; Menke 2007; and Isaac 1996.
8. For a helpful discussion on the public and public space in Arendt, see Benhabib 1993.
9. For a discussion of natality in Arendt, see Dietz 2010. Dietz outlines feminist debates over the usefulness of the concept of natality. See also Durst 2004; Guenther 2006; and Vatter 2006.
10. See Arendt1958a, 178.
11. See Arendt 1954, 61.
12. For important discussions of Arendt's notion of natality, see O'Byrne 2010; Birmingham 2006; Bowen-Moore 1989; Benhabib 2003.
13. For sustained discussion of Arendt's analysis of war and the importance of war in her thought, see Owens 2007. See also Birmingham 2010 and Bar On 2008.
14. Cf. Benhabib 2009, 331.
15. Butler relegates the distinction between earth and world to a mere footnote about Heidegger attached to a parenthetical remark about Arendt's "equivocation" on it (2012b 166, see also footnote 9 on page 238).
16. Wayne Allen discusses Arendt's notion of political imagination; see Allen 2002; see also Zerilli 2005. For engagements with Arendt and Kant on politics and imagination, see Beiner et al. 2001.
17. For insightful discussions of Arendt's notion of gratitude, see O'Byrne 2010 and Birmingham 2006. See also Chapman 2007 and Canovan 1994.
18. Peg Birmingham (2006) discusses Arendt's concept of the given.
19. See Arendt 1954, 263–264.
20. For helpful discussion of the concepts of earth and world alienation in Arendt's thought, see Canovan 1994; Chapman 2004, 2007; Macauley 1992, 1996; Ott 2009; Passerin d'Entreves 1991. Bonnie Mann develops an interesting analysis of Arendt's concept of world alienation in relation to feminism (Mann 2005).
21. Paul Ott describes the difference: "the products of labor are meant to be consumed, in the sense of entirely destroyed through consumption. They are perish-

able. Ideally, the waste from such activities should re-enter ecosystems with no net harmful results. The products of work, instead, are used up, but not destroyed all at once. They make up and are preserved in worlds" (Ott 2009, 14).

22. For discussions of consumption in Arendt's work, see Lulofs 1962; Mardellat 2011; and Norris 2005, 2006.
23. Ott gives some poignant examples of the ways that contemporary culture confuses labor and work (Ott, 2009, 14–15). And referring to Arendt's example of bread as a perishable consumable good as opposed to a table as a durable stable product, David Macauley points out that today some bread is made to last longer than some tables (1992, 40).
24. It would be interesting to consider innovation and newness in patent law in terms of Arendt's theory of natality as opening onto a future anterior tense, it will have been. In this regard, the new always comes to be through interpretation.
25. See Arendt 1954, 278.
26. George Kateb argues that both world and earth alienation are forms of homelessness caused by resentment towards our limitations and the limits of the human condition (1984, 164). "Restraint and self-restraint" he says, "would come from acceptance of the human condition and would lead to less alienation" (1984, 164). Whereas Kateb associates finding our way home with religion and spirituality, I would argue that for Arendt it comes with what she calls the "strange enterprise" of understanding, which, through imagination, can lead to shifting perspectives and thereby changing worlds.
27. See Macauley (1992, 26) for a discussion of planet as wanderer in relation to Arendt.
28. For a discussion of Arendt's *amor mundi,* see Chiba 1995. Chiba argues that Arendt's amor mundi needs to be supplemented by notions of eros and forgivness. (Chiba 1995, 509). See also Rose 1992; Miles 2002; and Martel 2008.
29. For sustained discussions of Arendt's notion of love, see also Beiner 1997.
30. See Arendt 1966 for a discussion of the right to rights. See also Honig 2006, 120.
31. Bonnie Honig reads Arendt's "right to have rights" in terms of Derrida's unconditional. (Honig 2006, 107).
32. In the words of Shin Chiba, Arendt's "notion of love can be seen as a principle for constituting a community, that is, a principle of coexistence—or life together—with whatever is outside and heterogeneous. . . . One's readiness to live together with those who are different, diversified, and heterogeneous, is the essential ingredient of amor mundi" (1995, 509 and 534).
33. Chiba argues that Arendt's notion of love the choice to live with others through friendship must be supplemented with Eros as the drive towards stable relationships with other people and the world (ibid., 509 and 534).
34. Cf. Arendt 1958a, 247.
35. For discussions of solidarity in Arendt's thought, see Allen 1999 and Reshaur 1992.

4. THE EARTH'S REFUSAL

1. For an insightful account of home in Heidegger's thought, see Capobianco 2005. For other discussions of home and homecoming, see Bambach 2009; Hammermeister 2000; Mugerauer 2008; and Tijmes 1998.
2. For an insightful discussion of *inhabit*, see Foltz 1995.
3. For a helpful discussion of *Being and Time*, see Kockelmans 1989. David Cerbone explains the role of World in Heidegger's early work (1995). See also Hall 1993. Klaus Held discusses Heidegger's notion of World in relation to Husserl (Held 1992).
4. There has been substantial secondary literature on Heidegger's concept of Dasein. On Dasein's gender neutrality, see Aho 2007; and, on Dasein in relation to race and ethnicity, see Ortega 2001, 2005. On the changing conception of Dasein, see Beistegui 2003. On Dasein in relation to community, see Boedeker 2001. On Dasein's authenticity, see Bracken 2005. On Dasein's moods, see Capobianco 2011. On Dasein's body, see Cerbone 2000 and Ciocan 2008. On Dasein in relation to Aristotle, see Hayes 2007. On Dasein and freedom, see Jaran 2010.
5. See Oliver 2009; for other commentaries on Heidegger and animals, see McNeill 1999; Franck 1991; Calarco 2004, 2008; Kuperus 2007. For even more creative readings of Heidegger's animal, see Derrida 1991, 2008 (chapter 4); Agamben 2003; Llewelyn 1991; Nancy 1997. Brett Buchanan also critically extends Heidegger's analysis of animals to make it more useful for thinking of reciprocity with animals (Buchanan 2008).
6. For an insightful discussion of Heidegger's notion of the uncanny in relation to Freud's, see Krell 1992.
7. For a sustained discussion of Heidegger's comparative analysis on the issue of hierarchy, see Oliver 2009.
8. John Caputo describes the relationship between boredom and refusal (1993, 54). For a discussion of the role of *Versagen* in *Being and Time*, see Marder 2007. Bernhard Radloff considers the role of refusal in *Contributions* (2007).
9. For sustained discussions on Heidegger's notion of boredom, see Biceaga 2006, Beistegui 2000; Emad 1985; Gibbs 2011; Gordon 2003; Hammer 2004 (here Espen Hammer discusses boredom in relation to melancholy), 2008; Mansikka 2009; McKenzie 2008; Slaby 2010; and Thiele 1997.
10. For extended discussions of Heidegger's position on animals in *The Fundamental Concepts*, see Oliver 2009. See also note 3, this chapter.
11. *Lassen* means both leaving and letting. "The Lassen is undecidable with respect to being transitive or intransitive: in this case it has the sense of both leaving something to be and making something be" (William McNeill and Nicholas Walker in Heidegger 1995 [1929–1930], 117). For an interesting discussion of Heidegger's notion of *Gelassenheit* in relation to Hölderlin, see Gosetti-Ferencei 2004. For a sustained engagement with Heidegger's *Gelassenheit*, see Davis 2007. See also George 2012.

12. There has been considerable commentary on Heidegger's "The Origin of the Work of Art." For example, Julian Young explores Heidegger's philosophy of art in this essay in relation to his corpus (Young 2004). See also Stulberg 1973.
13. Mitchell argues that this notion of strife undermines any sharp distinction between concealment and unconcealment and thereby destabilizes this opposition by presenting the earth as ungraspable sensuous appearing (Mitchell 2014, chapter 2).
14. For a comprehensive examination of Heidegger's notion of nature, see Foltz 1995.
15. See Mitchell 2014.
16. Cf. ibid., chapter 2.
17. Mitchell describes some of the differences between the notion of earth in *The Origin of the Work of Art* and the role of earth in the fourfold. See ibid.
18. For a discussion of the performative dimension of Heidegger's thought, especially in *Contributions*, see Wood 2002, chapter 10.
19. For example, while Michel Haar identifies earth with a "prehistorical ground," Robert Bernasconi describes earth as the native ground of a historical people.
20. Gregory Fried also associates earth with history and the strife between earth and world with *polemos* (the Greek word for *war* that Heidegger translates into German as *Auseinandersetzung* or confrontation, meaning everything from war to friendly debate) (Fried 2000, 15).
21. For the development of this argument, see Fried (ibid.). Although Fried occasionally equivocates earth and world (e.g., 63 and 75), his interpretation is illuminating.
22. Cf. Fried 2000, 61.
23. Ibid., 64–65.
24. For a discussion of Harris and other bioethicists who argue for genetic engineering as mastery, see Oliver 2013.
25. See Fried 2000, 66.
26. For example, whereas Iain Thomson identifies earth with a dimension of the intelligible, Fried claims that earth is not intelligible, and Frank Schalow describes the strife between earth and world as one between materiality and intelligibility (Fried 2000, 62; Schalow 2006, 94; Thomson 2011, 91).
27. Gadamer maintains, "The earth, in truth, is not stuff, but that out of which everything comes forth and into which everything disappears" (104). Both Haar and Schalow, on the contrary, associate earth with materiality (Haar 1993, 112; Schalow 2006, 94).
28. Similarly, Mitchell describes earth's materiality in terms of shining and sensuous appearance, particularly as it participates in the fourfold. (Mitchell 2014, chapter 2).
29. See Mølbak 2011, 218.
30. Gadamer stresses that Heidegger's concept of earth is a counterpoint to both Kantian subjective aesthetics and the objectivism of science (Gadamer 1994, 99 and 105). See also Benso 1997.
31. Engaging Heidegger and Derrida's commentaries on Heidegger, David Wood develops a notion of unlimited obligation. He also discusses the notions of limit and liminal in terms of Heidegger's thought (2002).

32. In "The Origin of the Work of Art" Heidegger discusses the *reliability of earth* in his description of Van Gogh's painting of the peasant woman's shoes. There Heidegger insists that the shoes are not just things or even less objects, but rather open onto the world of the peasant woman who uses them without thinking about them as shoes or instruments. It is their *reliability* that enables her to disregard them completely as she toils in the fields. Here too reliability is associated with the earth's telling refusal, its "silent call." The peasant woman hears the silent call of the earth through the reliability of her shoes, which also secures her world. In other words, the reliability of the shoes secures both the world and what disrupts the world, namely the refusal of the earth (1971c [1935–1936]).
33. Fried argues that interpretation is the medium of *polemos* (2000, 35).
34. *Introduction to Metaphysics* was published in 1935, the same year as "The Origin of the Work of Art," but first delivered as a lecture in 1931.
35. Bruce Foltz's *Inhabiting the Earth* is an excellent account of both the dangers and the saving power of technology (1995).
36. Economist Gao Shangquan says, "The difference of income per capita between the richest country and poorest country has enlarged from 30 times in 1960 to the current 70 times" (Shangquan 2000, 4).
37. For an insightful discussion of the planetary at odds with the earth in Heidegger, see Turnbull 2006.
38. For an excellent analysis of Heidegger's changing attitude toward home, see Richard Capobianco 2005.
39. Quoted in Capobianco 2005, 164.
40. Foltz Bruce says, "The earth as a 'homeland' does not refer to some sort of atavistic and reactionary political agenda, as some have maintained, but simply to the 'nearness' and 'significance' of nature, which are jeopardized by modern science and technology" (1995, 144).
41. Already in *Contributions*, we see the fore-figures of the fourfold. What in *Contributions* appear as gods and humans, earth and world, become transformed into gods and mortals, earth and sky as the four elements of the fourfold (*das Geviert*). See Heidegger 2012b, §268, 377 and §270, 381.
42. For a sustained discussion of the role of earth in Heidegger's fourfold, see Mitchell 2014.
43. See Young 2000, 373.
44. There are radically different interpretations of earth in the fourfold. For example, compare Young 2000 to Mitchell 2009.
45. Mitchell calls earth the "groundless ground" (2009, 212); see also Mitchell 2014.
46. Thanks to Andrew Mitchell for pointing out that while in *Being and Time* Dasein is given, in *Contributions* it is an achievement of man.
47. For an argument along these lines, see Wood 2012.
48. For a sustained analysis of Heidegger's comparative pedagogy, see Oliver 2009.
49. While Young sees sky as the "literal sky" (2006, 374), along with what comes from it including weather, seasons, day and night, and planetary motions, Mitchell in-

terprets these elements as the "space of the earth's emergence, the space wherein things appear and through which they shine" (2009, 213).
50. Whereas Young gives a fairly traditional reading of humans as mortals because they are "capable of death as death" and unlike animals they are "capable of approaching death with an understanding of what it truly is," Mitchell gives a more subtle interpretation of death in *Being and Time* that is resonant with Heidegger's late work (see Young 2006, 375). Mitchell argues that although in *Being and Time* death is our "own," we do not possess it, but rather we are dispossessed by it (2009, 211).
51. See Mitchell 2009 for an analysis of the dispossession inherent in each element of the fourfold.
52. See Mitchell (2009, 211).
53. Mitchell argues that Young's interpretation of divinities as "heroes" or "mythologized figures preserved in the collective memory of a culture" misses their role in a hermeneutics of the holy (Mitchell 2009, 217n16); see Young 2006, 374.
54. Mitchell argues that Heidegger's divinities (*die Göttlichen*) are the messengers of the holy (2009, 214).
55. Ben Vedder analyzes Heidegger's concept of the Holy in relation to the Whole by way of his concepts of anxiety, boredom and wonder (Vedder 2005).
56. See Heidegger 1971c [1935–1936], 170 and 1971a [1951], 355.
57. Cf. Heidegger 1966, 108.
58. On community and belonging in Heidegger, see Birmingham 1991 and Odysseos 2009, both of whom are responding to Philippe Lacoue-Labarthe's critical assessment of Heidegger's politics (Lacoue-Labarthe 1990).
59. See Mitchell 2009, 215.
60. For a sustained discussion of ethical responsibility as enabling the ability to respond, see Oliver 2001.
61. See Smith 2011 for an attempt to extend Heidegger's analysis of the saving power as letting be.
62. Cf. Mitchell 2011, 8. Mitchell argues that Heidegger emphasizes the connection between ground and earth as the nourishing soil (2011, 9).
63. See Lazier 2011, 609–610.
64. Mikko Joronen argues that globalization is an instantiation of technological enframing (Joronen 2008). See also Lazier 2011.
65. In a Heideggerian vein, Jean-Luc Nancy comments on the satellites encircling the earth (Nancy 2007, 33).
66. Cf. Andrew Mitchell on unworlding (Mitchell 2005, 197).
67. See Derrida's *Of Spirit* for a discussion of "only a god can save us now" (1989).
68. The German word is *Not,* which does not have the connotation of vow or pledge, but does connote emergency and crisis.
69. Mitchell describes what Heidegger means by shining; see Mitchell 2014, chapter 2.
70. For an interesting analysis of Heidegger's move from angst in his earlier work to astonishment in his later work, see Capobianco 2011.

71. For an insightful analysis of Heidegger's notion of poetic dwelling in relationship to the Earth and environmentalism see Foltz 1995. There Foltz offers one of the most developed accounts of how Heidegger's thoughts about nature and earth can contribute to environmental philosophy (176). Mick Smith also uses some of Heidegger's philosophy of earth and world in order to argue for an environmentalism of letting be (Smith 2011). Other contributions to environmental philosophy that use resources from Heidegger include Brown and Toadvine 2003; Haar 1993; Irwin 2008; McWhorter and Stenstad 1992; Schalow 2006; and Zimmerman 1983, 1993, and 1994. Irwin 2008 extends Heidegger's concepts of nature and technology to discuss climate change.

5. THE WORLD IS NOT ENOUGH

1. For a discussion of Arendt's notion of the right to rights in terms of Derrida's notion of hospitality, see Honig 2006.
2. Andrew Mitchell calls Heidegger's notion of Earth a "groundless ground" (2009, 212).
3. Honig discusses Arendt's notion of the right to have rights in terms of Derrida's notion of hospitality (2006, 107).
4. For insightful commentaries on various aspects of Derrida's *The Beast and the Sovereign, Volume II*, see Chin-Yi 2012; Harafin 2013; Krell 2013; and Naas 2012a, 2014. Krell and Naas both offer insightful and sustained analyses of the entire seminar. Naas's chapter on world, "If you could take just two books," deals with many of the same themes that I address here. And I am indebted to him for allowing me to read his manuscript before its publication. David Krell's *Derrida and Our Animal Others* is an interesting combination of straightforward exegesis of the seminars in chronological order and a defense of Heidegger against some of Derrida's "perverse readings," all combined with Krell's own rhetorical flourish.
5. I have discussed Derrida's argument in *The Animal That Therefore I Am* elsewhere, see Oliver 2009. For insightful readings of this text, see Lawlor 2007; Calarco 2008; Wolfe 2008. See also Wood 1999, 2004; Naas 2013; and Krell 2013 for discussions of Derrida on animals.
6. See Derrida 2008, 1989, and the Geschlecht essays (1993, 1991, 1987).
7. Heidegger (1995 [1929–1930]) also mentions death in passing or as an example of other's misinterpretation of *Being and Time* (1996a [1927]) on death on pages 26, 61, 265, 267, 273, 294–297, 300. On pages 294–297, he summarizes his notion from *Being and Time* that Dasein is being toward death and responds to misinterpretations of it.
8. At the end of chapter 5, Heidegger uses death to make the point that motility, like death, is essential to understanding life as such (Heidegger 1995 [1929–1930], 266). He goes on to take up the animal in relation to death; referring back to his discussion of the animal's captivation in its disinhibiting ring: "Is the death of the animal a dying or way of coming to an end?" (267). As we know, he answers this question

with the famous claim that animals do not die as such, but merely come to an end, echoing remarks that he makes elsewhere (e.g., "The Nature of Language" in Heidegger 1971b [1950–1959], 107). Just as quickly as he turned to death, however, he leaves it behind and returns to the question of the essential nature of the thesis that the animal is poor in world (Heidegger 1995 [1929–1930], 267). The only other noteworthy mention of death in *Fundamental Concepts* is at §70, where Heidegger summarizes his notion of Dasein's being-toward-death in *Being and Time*, but again only to make a related point about the difference between philosophy and science (Heidegger 1995 [1929–1930], 294–295).

9. It is noteworthy that Derrida uses the fragment from the Celan poem at moments when he shifts his attention from Heidegger's *Fundamental Concepts of Metaphysics* to other texts by Heidegger, specifically *The Introduction to Metaphysics* and *Identity and Difference,* both in session 10. We might say, then, that this Celan fragment operates as a type of *metaphora* in the sense that Derrida invokes it in *The Beast and the Sovereign* as source of movement or, in other terms, displacement.

10. Bernasconi 2000 and Cohen 2006 offer fascinating discussions of Heidegger and Levinas on being toward death and the death of the other. Iain Thomson engages Heidegger and Derrida on the question of "Can I die?" (Thomson 1999).

11. For other discussions of Derrida's commentaries on the poetry of Paul Celan, see Fioretos 1990 and Pasanen 2006.

12. This sense of déjà vu might be the result of Derrida's earlier engagements with Celan, particularly with this poem, in *Rams* (2005b).

13. The poem can be found in Paul Celan's 1967 collection *Atmenwende (Breathturn)* (Celan 1986 [1967]).

14. Cf. Heidegger 1995, 269–270.

15. Derrida comments on Robinson Crusoe's terror at not knowing whose footprint it could be, that of another or his own (Derrida 2011, 48).

16. For a discussion of the prosthesis of the world through which the individual becomes unique, see Naas 2014.

17. For a discussion of prayer in *The Beast and the Sovereign, Volume II*, see Naas 2014, chapter 5.

18. See ibid., 49–50.

19. Michael Naas discusses this passage at length (ibid., 52–54).

20. See ibid., 52.

21. My translation in text.

22. My translation in text.

23. For discussions of the Derridean "as if" see Dickinson 2011; Fujita 2012; and Naas 2008.

24. Derrida sets out the *both-and* aspect of whether or not we share a world in the following two long sentences, see Derrida 2011, 265–266.

25. For a discussion of poetic majesty as it operates in *The Beast and the Sovereign, Volume I*, see Oliver 2013, chapter 3. For a discussion of *poesis* in Derrida's ethics and the "mechanics of deconstruction," see Hansen 2000, especially chapter 5.

26. See Derrida 2009, 273. It is noteworthy that Derrida's reference to breath in this passage is an allusion to Celan's poem "Meridian," which he discusses at length in the seminar.
27. For discussions of hyperbolic ethics, see Attridge 2007; Llewelyn 2002; Marder 2010; Rottenberg 2006; Weber 2005; Zlomislić 2007; and Oliver 2009, 2013.
28. For a discussion of Derrida's claim in "No Apocalypse, Not Now" that "there is no common measure able to persuade me that a personal mourning is less grave than a nuclear war," see Naas 2014, 51–52.
29. For discussions of Derrida's relationship to the Kantian as if, see Naas 2008; Dickinson 2011; and Fujita 2012.
30. For an insightful discussion of the relationship of Derrida's use of as if to the *as if* and *as such* of philosophy, especially Plato, see Naas 2008, especially pages 15, 37–38, 45, 53, 79, 188, 200, 207, 238n6.
31. For a discussion of Derrida's impossible ethics, see Raffoul 2008.
32. In his Bremen acceptance speech, Celan describes language and poetry: "Only one thing remained reachable, close and secure amid all losses: language.... A poem, being an instance of language, hence essentially dialogue, may be a letter in a bottle thrown out to sea with the—surely not always strong—hope that it may somehow wash up somewhere, perhaps on a shoreline of the heart" (1986, 35).
33. E.g., Gadamer 1983; Felman and Laub 1992.
34. Derrida is quoting Hölderlin's poem "Die Titanen."
35. Many species of nonhuman animals reportedly acknowledge, mourn, and even bury their dead. Different species of birds, magpies and scrub jays, have been seen engaging in ritualistic behavior around their dead (Bekoff 2009). Sea Lions wail in mourning at the death of their young or mates (Bender 2012). Normally active and boisterous chimpanzees give a moment of silence to the dead (Hanlon 2009). Gorillas and dolphins also exhibit ritualistic or unusual behaviors around the death of their companions (Bender 2012). Geese, llamas, and wolves also mourn their dead (Bekoff 2009).
36. Kirby 2012. See also Waterworth 2012.
37. See Bradshaw and Shore 2007; O'Connell-Rodwell 2011; McComb et al. 2001; Highfield 2006; Bradshaw 2004; Bekoff 2009; and Siebert 2011.

6. TERRAPHILIA

1. See Derrida 2009, 15–16. For a discussion of Derrida as a philosopher of limits, see Oliver 2013.
2. For a helpful discussion of the tension between the ethical and the political in Derrida's work in relation to Kant, see La Caze 2007; see also Oliver 2012. For helpful discussion of Derrida's notion of justice, see Cornell 2006; Cornell, Rosenfeld, and Carlson 1992; Fritsch 2011; Goodrich 2008; Honig 1991 (Honig compares Arendt and Derrida); Jennings 2006; Naas 2005; and Weber 1989.

3. John Sallis argues, "Having the earth in common with all men does not prescribe sharing any part of that surface with any who might occupy it; it does not prescribe any rule of hospitality. Having the earth in common does not, as such, produce any coherence among men, any substantial or essential community" (Sallis 1998, 207). As he points out, to spite sharing the limited surface of the earth, we still have wars and violence.
4. Cf. ibid.
5. For a discussion of "thickening" in relation to Derrida and Rousseau, see my *Animal Lessons,* Oliver 2010.
6. Cf. Sallis 1998, 208. Following Heidegger, insofar as the earth both shelters and withdraws, Sallis concludes, through "the earth itself a community in withdrawal would (have) come to play" (ibid.).
7. For alternative attempts to formulate notions of community that start with singularity, see Jean-Luc Nancy 1991, 2000. For excellent engagements with Nancy's notion of community, see O'Byrne 2010 and Schroeder 2004. Brian Schroeder uses Nancy's notion of the *inoperative community* to develop a notion of what he calls *inoperative earth*.
8. In *The Time of Life*, William McNeill explores Heidegger's ethics in terms of ethos and argues for ethics as a way of life rather than moral rules or principles (McNeill 2007).
9. For a provocative use of Derridean deconstruction to diagnose the current situation of environmental philosophy, particularly the situation where we must decide and act before we have proof of climate change, see Wood 2006.
10. See Rifkin 2009 for a survey of literature on so-called selfish genes and altruism as "hard-wired" into human infants and other species.
11. For a discussion of the importance of play in the development of empathy in humans and other animals, see Rifkin 2009. See also Pellegrini et al. 2007. For a discussion of some of these studies in relation to philosophy, see Willett 2014. Rifkin cites studies on horses and play, particularly Overton and Doods 2006. See also Vygotsky 1978, also discussed in Rifkin 2009.
12. Among many other recent developments, he appeals to the discovery of what neuroscientists call "mirror neurons" or "empathy neurons" in humans and higher primates as proof that we are empathic "by nature" and that our empathic sensibilities are evolving (2009, 83–87).
13. Rifkin concludes: "If we can harness our empathetic sensibility to establish a new global ethic that recognizes and acts to harmonize the many relationships that make up the life-sustaining forces of the planet, we will have moved beyond the detached, self-interested and utilitarian philosophical assumptions that accompanied national markets and nation state governance and into a new era of biosphere consciousness" (ibid.).
14. See ibid., chapter 11. See also Davi 2014; and the Harris Poll 2007.
15. For an assessment of diminishing biodiversity, see Wilson 1993.

16. Along with subjectless sociality, intersubjective attunement, and spirituality and compassion, Willett identifies the biosocial network as home as one of what she calls the four layers of interspecies ethics (Willett 2014, 133).
17. Willett concludes: "If we had to identify some final meaning for ethics, its principleless principle would not be found in philosophical *logos* but in the playful encounter.... Affects and emotions shape the biosocial drives and desires (*Eros*) for friendly bonds and a sense of home" (ibid., 134).
18. Willett says, "*Eros* is not a bare striving for pleasure or wild intensity but a meaning-laden yearning. *Eros* is a drive toward home" (ibid., 23).
19. Maurice Merleau-Ponty discusses animal cultures in his *Nature Lectures* (2003). For an analysis of Merleau Ponty on animal culture, see Oliver 2009.
20. Further complicating any easy and comfortable notion of home, from a political perspective, home can be a contested space, sometimes filled with violence.
21. Compare Cynthia Willett's discussion of home in *Interspecies Ethics* (2014). See also Rifkin 2009. There, he summarizes psychological studies that indicate that the infant's drive to belong is primary (ibid., 20–21). Some of the psychologists who emphasize attachment and belonging are Kohut, Winnicott, and Bowlby. In her *Maternal Ethics*, Cynthia Willett also discusses some of these attachment theorists, particularly in relation to French feminism and more contemporary developments (Willett 1995).
22. Developing a Heideggerian analysis of terror and terrorism, Andrew Mitchell calls "homeland security" an oxymoron; see Mitchell 2005.
23. Throughout her work, Cynthia Willett has argued for a relational notion of freedom. See, e.g., Willett 2008, 124.
24. See Willett's discussion of home in the introduction to her recent book, *Interspecies Ethics* (2014).
25. For a nice discussion of the relationship between rethinking earth and community, see Schroeder 2004.
26. Cf. Willett 2014, chapter 5, 131–146.
27. My colleague David Wood often says that instead of missions to the moon or to Mars, we need a mission to Earth.
28. For example, Warsaw Climate Change Summit 2013, participants: 195 countries; G-20 Climate Summit, G-20 Major Economies; Doha Climate Change Summit 2012, participants: 195 countries; Durban Climate Change Summit 2011, participants: 194 countries; Cancún Climate Change Summit 2010, participants: 194 countries; Copenhagen Climate Change Summit 2009, participants: 194 countries.
29. This tension is repeated in more contemporary debates over nationalism versus cosmopolitanism, especially post-Kantian philosophies that attempt to embrace both. I am thinking of Seyla Benhabib in particular. She makes the tensions explicit in *Another Cosmopolitanism* (Benhabib 2006).

BIBLIOGRAPHY

Abbott, Matthew. 2010. "The Poetic Experience of the World." *International Journal of Philosophical Studies* 18 (4): 493–516.
Abram, David. 1988. "Merleau-Ponty and the Voice of the Earth." *Environmental Ethics* 10 (2): 101–120.
Agamben, Giorgio. 2003. *The Open: Man and Animal.* Trans. Kevin Attell. Stanford: Stanford University Press.
Aldrin, Buzz. 2013. "The Call of Mars." *New York Times*, June 13. http://www.nytimes.com/2013/06/14/opinion/global/buzz-aldrin-the-call-of-mars.html.
Allen, Amy. 1999. "Solidarity After Identity Politics: Hannah Arendt and the Power of Feminist Theory." *Philosophy and Social Criticism* 25 (1): 97–118.
Allen, Wayne. 2002. "Hannah Arendt and the Political Imagination." *International Philosophical Quarterly* 42 (3): 349–369.
Aho, Kevin. 2007. "Gender and Time: Revisiting the Question of Dasein's Neutrality." *Epoch* 12 (1): 137–155.
Arendt, Hannah. 1953. "Understanding and Politics." *Partisan Review* 20:377–392.
———. 1954. Between Past and Future. New York: Penguin.
———. 1958a. *The Human Condition.* Chicago: University of Chicago Press.
———. 1958b. "The Modern Concept of History." *Review of Politics* 20 (4): 570–590.
———. 1959. "Reflections on Little Rock." *Dissent* 6 (1) (Winter): 45–55.
———. 1960. "Freedom and Politics: A Lecture." *Chicago Review* 14 (1): 38–46.
———. 1966. *The Origins of Totalitarianism,* New York: Harcourt, Brace, and World.
———. 1968a. "Karl Jaspers: Citizen of the World." In *Men in Dark Times,* 81–94. New York: Harcourt, Brace, and World.

———. 1968b. *Totalitarianism: Part Three of the Origins of Totalitarianism.* New York: Houghton Mifflin Harcourt.
———. 1970. *On Violence.* New York: Harcourt, Brace, and World.
———. 1971. "Thinking and Moral Considerations: A Lecture." *Social Research* 38 (3): 417–446.
———. 1975. "Home to Roost: A Bicentennial Address." *New York Review of Books* 22 (11) (June 26).
———. 1981. *The Life of the Mind.* New York: Harcourt.
———. 1992. *Eichmann in Jerusalem: A Report on the Banality of Evil.* New York: Penguin.
———. 2005. *The Promise of Politics.* Ed. Jerome Kohn. New York: Schocken.
———. 2007. "The Conquest of Space and the Stature of Man." *New Atlantis* (Fall): 43–55.
Arendt, Hannah, and Karl Jaspers. 1993. *Briefwechsel Arendt / Jaspers 1926 –1969,* ed. Lotte Köhler and Hans Saner. *Sonderausgabe.* Munich: Piper.
Ataner, Attila. 2006. "Kant on Capital Punishment and Suicide." *Kant-Studien* 97 (4): 452–482.
Attridge, Derek. 2007. "The Art of the Impossible?" In *The Politics of Deconstruction: Jacques Derrida and the Other of Philosophy.* Ed. Martin McQuillan. London: Pluto.
Avise, John C. 1994. "Editorial: The Real Message from Biosphere 2." *Conservation Biology* 8 (2): 327–329.
Babich, Babette. 2003. "From Van Gogh's Museum to the Temple at Bassae: Heidegger's Truth of Art and Schapiro's Art History." *Culture, Theory, and Critique* 44 (2): 151–169.
Bambach, Charles. 2009. "Situating Heidegger." *American Catholic Philosophical Quarterly* 83 (4): 599–613.
Bartky, Sandra Lee. 1979. "Heidegger and the Modes of World-Disclosure." *Philosophy and Phenomenological Research* 40 (2): 212–236.
Bar On, Bat-Ami. 2008. "The Opposition of Politics and War." *Hypatia* 23 (2): 141–154.
Bay, Jennifer and Thomas Rickert. 2008. "New Media and the Fourfold." *JAC* 28 (1/2): 209–244.
Beiner, Ronald. 1997. "Love and Worldliness: Hannah Arendt's Reading of Saint Augustine." In *Hannah Arendt: Twenty Years Later,* ed. Larry May and Jerome Kohn. Cambridge: MIT Press.
Beiner, Ronald, and Jennifer Nedelsky, ed. 2001. *Judgment, Imagination, and Politics: Themes from Kant and Arendt.* Lanham, MD: Rowman and Littlefield.
Beistegui, Miguel de. 2000. "'Boredom: Between Existence and History.' On Heidegger's Pivotal *The Fundamental Concepts of Metaphysics.*" *Journal of the British Society for Phenomenology* 31 (2): 145–158.
———. 2003. "The Transformation of the Sense of Dasein in Heidegger's *Beiträge zur Philosophie (Vom Ereignis)*." *Research in Phenomenology* 33 (1): 221–246.
Bekoff, Marc. 2009. "Grief in Animals: It's Arrogant to Think We're the Only Animals Who Mourn." *Psychology Today,* October 29. www.psychologytoday.com/blog/

animal-emotions/200910/grief-in-animals-its-arrogant-think-were-the-only-animals-who-mourn.

Bender, Kelli. 2012. "10 Heartbreaking Animal Mourning Rituals." *PawNation*. www.pawnation.com/2012/07/31/10-heartbreaking-animal-mourning-rituals/.

Bendle, Mervyn F. 2005. "The Apocalyptic Imagination and Popular Culture." *Journal of Religion and Popular Culture* 11 (Fall): online.

Benhabib, Seyla. 1993. "Feminist Theory and Hannah Arendt's Concept of Public Space."*History of the Human Sciences* 6 (2): 97–114.

——. 2003. *The Reluctant Modernism of Hannah Arendt*. Lanham, Md.: Rowman and Littlefield.

——. 2006. *Another Cosmopolitanism*. New York: Oxford University Press.

——. 2007. "Another Universalism: On the Unity and Diversity of Human Rights." *Proceedings and Addresses of the American Philosophical Association* 81 (2): 7–32.

——. 2009. "International Law and Human Plurality in the Shadow of Totalitarianism: Hannah Arendt and Raphael Lemkin." *Constellations* 16 (2): 331–350.

——. 2010. "Introduction." In *Politics in Dark Times: Encounters with Hannah Arendt*, ed. Seyla Benhabib. New York: Cambridge University Press.

Benso, Silvia. 1997. "The Face of Things: Heidegger and the Alterity of the Fourfold." *Symposium* 1 (1): 5–15.

Berger, John. 1991. *About Looking*. New York: Random House.

Bernasconi, Robert. 1989. "History Is Seldom": Holderlin and Heidegger. In *The Question of Language in Heidegger's History of Being*, 29–47. Atlantic Highlands, NJ: Humanities.

——. 1996. *Heidegger in Question*. New York: Humanity.

——. 2000. "Whose Death Is It Anyway? Philosophy and the Cultures of Death." *Tympanum: A Journal of Comparative Literary Studies* 4. www.usc.edu/dept/comp-lit/tympanum/4/bernasconi.html.

——. 2010. "Race and Earth in Heidegger's Thinking During the Late 1930s." *Southern Journal of Philosophy* 4 (1): 49–66.

Bernstein, Susan. 2003. "It Walks: The Ambulatory Uncanny." *MLN* 118 (5): 1111–1139.

Biceaga, Victor. 2006. "Temporality and Boredom." *Continental Philosophy Review* 39 (2): 135–153.

"Biosphere 2." *UA Science Connections*. http://scienceconnections.arizona.edu/community/venues/biosphere2. Last Accessed July 12, 2013.

Birmingham, Peg. 1991. "The Time of the Political." *Graduate Faculty Philosophy Journal* 14–15 (2–1): 25–45.

——. 2006. *Hannah Arendt and Human Rights: The Predicament of Common Responsibility*. Bloomington: Indiana University Press.

——. 2010. "On Violence, Politics, and the Law." *Journal of Speculative Philosophy* 24 (1): 1–20.

Bishop, Peter. 1986. "The Shadows of the Holistic Earth." *Spring Journal*, pp. 59–71.

Biskowski, Lawrence J. 1993. "Practical Foundations for Political Judgment: Arendt on Action and World." *Journal of Politics* 55 (4): 867–887.

Blumenberg, Hans. 1989. *The Genesis of the Copernican World*. Cambridge: MIT Press.
———. 1996. *Shipwreck with Spectator: Paradigm of a Metaphor for Existence*. Cambridge: MIT Press.
Boedeker, Edgar C. 2001. "Individual and Community in Early Heidegger: Situating Das Man, the Man-Self, and Self-Ownership in Dasein's Ontological Structure." *Inquiry* 44 (1): 63–99.
Bonta, Mark and John Protevi. 2004. *Deleuze and Geophilosophy: A Guide and Glossary*. Edinburgh: Edinburgh University Press.
Bowen-Moore, Patricia. 1989. *Hannah Arendt's Philosophy of Natality*. Macmillan.
Bowman, Curtis. 2003. "Heidegger, the Uncanny, and Jacques Tourner's Horror Films." In *Dark Thoughts: Philosophic Reflections on Cinematic Horror*, ed. Steven Jay Schneider and Daniel Shaw, 65–83. Lanham, MD: Scarecrow.
Boym, Svetlana. 2009. "From Love to Worldliness: Hannah Arendt and Martin Heidegger." *The Yearbook of Comparative Literature* 55:106–128.
Bracken, William F. 2005. "Is There a Puzzle About How Authentic Dasein Can Act? A Critique of Dreyfus and Rubin on Being and Time, Division II." *Inquiry* 48 (6): 533–552.
Bradshaw, Isabel G. A. 2004. "Not by Bread Alone: Symbolic Loss, Trauma, and Recovery in Elephant Communities." *Society and Animals* 12, no. 2: 143–158.
Bradshaw, Isabel G. A. and Allan Schore. 2007. "How Elephants Are Opening Doors: Developmental Neuroethology, Attachment, and Social Context." *Ethology* 113:426–436.
Brannigan, John, Ruth Robbins, and Julian Wolfreys, eds. 1996. "As If I Were Dead: An Interview with Jacques Derrida." In *Applying: To Derrida*, 212–226. New York: St. Martin's.
Broad, William J. 1996. "Paradise Lost: Biosphere Retooled as Atmospheric Nightmare." *New York Times*, November 19. www.nytimes.com/1996/11/19/science/paradise-lost-biosphere-retooled-as-atmospheric-nightmare.html. Last Accessed July 12, 2013.
Broderick, Mick. 1991. *Nuclear Movies: A Critical Analysis and Filmography of International Feature-Length Films Dealing with Experimentation, Aliens, Terrorism, Holocaust, and Other Disaster Scenarios, 1914–1990*. Jefferson, NC: McFarland.
Brown, Charles and Ted Toadvine, eds. 2003. *Eco-phenomenology: Back to the Earth Itself*. Albany: SUNY Press.
Brown, Garrett W. 2010. "The Laws of Hospitality, Asylum Seekers and Cosmopolitan Right: A Kantian Response to Jacques Derrida." *European Journal of Political Theory* 9 (3): 308–327.
Bruin, John. 1992. "Heidegger and the World of the Work of Art." *Journal of Aesthetics and Art Criticism* 50 (1): 55–56.
———. 1994. "Heidegger and Two Kinds of Art." *Journal of Aesthetics and Art Criticism* 52 (4): 447–457.
Brunkhorst, Hauke. 1996. "Are Human Rights Self-Contradictory?" Critical Remarks on a Hypothesis by Hannah Arendt." *Constellations* 3 (2): 190–199.

Bryan, Bradley. 2012. "Revenge and Nostalgia: Reconciling Nietzsche and Heidegger on the Question of Coming to Terms with the Past." *Philosophy and Social Criticism* 38 (1): 25–38.

Buchanan, Brett. 2008. *Onto-Ethologies: The Animal Environments of Uexkull, Heidegger, Merleau-Ponty, and Deleuze*. Albany: SUNY.

———. 2012. "Being with Animals: Reconsidering Heidegger's Animal Ontology." In *Animals and the Human Imagination: A Companion to Animal Studies*, ed. Aaron Gross and Anne Vallely, 265–288. New York: Columbia University Press.

Butler, Judith. 2011. "Hannah Arendt's Death Sentences." *Comparative Literature Studies* 48 (3): 280–295.

———. 2012a. "Precarious Life, Vulnerability, and the Ethics of Cohabitation." *Journal of Speculative Philosophy* 26 (2): 134–151.

———. 2012b. *Parting Ways: Jewishness and the Critique of Zionism*. New York: Columbia University Press.

Butler, Judith and Gayatri Spivak. 2011. *Who Sings the Nation-State? Language, Politics, Belonging*. Chicago: Seagull.

Burke, Anthony. 2008. "Recovering Humanity from Man: Hannah Arendt's Troubled Cosmopolitanism." *International Politics* 45 (4): 514–521.

Byrd, B. Sharon and Joachim Hruschka. 2010. "Intelligible Possession of Land." In *Kant's Doctrine of Right: A Commentary*, 122–142. Cambridge: Cambridge University Press.

Calarco, Matthew. 2004. "Heidegger's Zoontology." In *Animal Philosophy*, ed. Peter Atterton and Matthew Calarco, 18–30. New York: Bloomsbury Academic.

———.2008. *Zoographies: The Question of the Animal from Heidegger to Derrida*. New York: Columbia University Press.

Calder-Williams, Evan. 2011. *Combined and Uneven Apocalypse: Luciferian Marxism*. Hampshire: Zero.

Canovan, Margaret. 1983. "Arendt, Rousseau, and Human Plurality in Politics." *Journal of Politics* 45 (2): 285–302.

———. 1994. *Hannah Arendt: A Reinterpretation of Her Political Thought*. Cambridge: Cambridge University Press.

Capobianco, Richard. 2005. "Heidegger's Turn Toward Home: On Dasein's Primordial Relation to Being." *Epoché* 10 (1): 155–173.

———. 2011. "From Angst to Astonishment." In *Engaging Heidegger*, 70–86. Toronto: University of Toronto Press.

Caputo, John D. 1988. *Radical Hermeneutics: Repetition, Deconstruction, and the Hermeneutic Project*. Bloomington: Indiana University Press.

———. 1993. *Demythologizing Heidegger*. Bloomington: Indiana University Press.

Carlson, Thomas A. 2007. "With the World at Heart: Reading Cormac McCarthy's *The Road* with Augustine and Heidegger." *Religion and Literature* 39 (3): 47–71.

Casey, Edward S. 2001. "Between Geography and Philosophy: What Does It Mean to Be in the Place-World?" *Annals of the Association of American Geographers* 91 (4): 683–693.

Celan, Paul. 1986. *Collected Prose*. Trans. Rosmarie Waldrop. Manchester: Carcanet.
———. 1995. *Breathturn*. Los Angeles: Sun and Moon.
Cerbone, David R. 1995. "World, World-Entry, and Realism in Early Heidegger." *Inquiry* 38 (4): 401–421.
———. 2000. "Heidegger and Dasein's 'Bodily Nature': What Is the Hidden Problematic?" *International Journal of Philosophical Studies* 8 (2): 209–230.
Chambers, Iain. 2001. "Earth Frames: Heidegger, Humanism and 'Home.'" In *Culture After Humanism: History, Culture, Subjectivity*, 47–74. New York: Routledge.
Chapman, Anne. 2004. "Technology as World Building." *Ethics, Place, and Environment* 7 (1–2): 59–72.
———. 2007. "The Ways That Nature Matters: The World and the Earth in the Thought of Hannah Arendt." *Environmental Values* 16 (4): 433–445.
Cheah, Pheng. 2009. "The Untimely Secret of Democracy." In *Derrida and the Time of the Political*, ed. Pheng Cheah and Suzanne Guerlac, 74–96. Durham: Duke University Press.
Chiba, Shin. 1995. "Hannah Arendt on Love and the Political: Love, Friendship, and Citizenship." *Review of Politics* 57 (3): 505–535.
Chin-Yi, Chung. 2012. "The Opposition Between Theism and Atheism in Derrida's *The Beast and the Sovereign, Volume II*." *Quint* 4 (4): 95–98.
Cholbi, Michael. 2000. "Kant and the Irrationality of Suicide." *History of Philosophy Quarterly* 17 (2): 159–176.
"Chronology of Biosphere 2." *Bio Spherics*. www.biospherics.org/biosphere2/chronology/. Last accessed July 12, 2013.
Ciocan, Cristian. 2008. "The Question of the Living Body in Heidegger's Analytic of Dasein." *Research in Phenomenology* 38 (1): 72–89.
Cixous, Helene. 2004. "Birds, Women, and Writing." In *Animal Philosophy: Ethics and Identity*, ed. Peter Atterton and Matthew Calarco, 167–173. London: Continuum.
———. 2008. "But the Earth Still Turns, and Not as Badly as All That." In *White Ink: Interviews on Sex, Text, and Politics*, ed. Susan Sellers, 180–182. New York: Columbia University Press.
Cixous, Helene, Catherine Anne Franke, and Roger Chazal. 1989. "Interview with Helene Cixous." *Qui Parle* 3 (1): 152–179.
Clark, David. 2001. "Kant's Aliens: The Anthropology and Its Others." *New Centennial Review* 1 (2): 201–289.
Cohen, Alix A. 2006. "Kant on Epigenesist, Monogenesis and Human Nature: The Biological Premises of Anthropology." *Studies in History and Philosophy of Biological and Biomedical Sciences* 37:675–693.
———. 2008. "Kant's Biological Conception of History." *Journal of the Philosophy of History* 2 (1): 1–28.
Cohen, Richard A. 2006. "Levinas: Thinking Least About Death: Contra Heidegger." *International Journal for Philosophy of Religion* 60 (1/3): 21–39.
Colebrook, Claire. 2012. "A Globe of One's Own: In Praise of the Flat Earth." *SubStance* 41 (1): 30–39.

Colony, Tracy. 2007. "Before the Abyss: Agamben on Heidegger and the Living." *Continental Philosophy Review* 40:1–16.
Conley, Verena Andermatt. 1997. *Ecopolitics: The Environment in Poststructuralist Thought*. New York: Routledge.
Cornell, Drucilla. 2006. "Rethinking Legal Ideals After Deconstruction." In *Law's Madness*, ed. Austin Sarat, Lawrence Douglas, and Martha Merrill Umphrey, 147–168. Ann Arbor: University of Michigan Press.
Cornell, Drucilla, Michel Rosenfeld, and David Gray Carlson, eds. 1992. *Deconstruction and the Possibility of Justice*. New York: Routledge.
Cosgrove, Denis. 1994. "Contested Global Visions: One-World, Whole-Earth, and the Apollo Space Photographs." *Annals of the Association of American Geographers* 84 (2): 270–294.
———. 2003. *Apollo's Eye: A Cartographic Genealogy of the Earth in the Western Imagination*. Baltimore: Johns Hopkins University Press.
Crang, Mike and N. J. Thrift, eds. 2000. *Thinking Space*. New York: Routledge.
Crockett, Clayton. 2001. "On God and Being: A Review of Martin Heidegger's *Contributions to Philosophy*." *Journal for Cultural and Religious Theory* 2 (2).
———. 2003. "Foreclosing God: Philosophy of Religion and Psychoanalysis." In *Explorations in Contemporary Continental Philosophy of Religion*, ed. Deane-Peter Baker and Patrick Maxwell, 175–188. New York: Rodopi.
Darwin, Charles. 1981 [1871]. *The Descent of Man, and Selection in Relation to Sex*. Princeton: Princeton University Press.
Dastur, Francoise. 2000. *Telling Time: Sketch of a Phenomenological Chronology*. New York: Continuum.
Davi, Robert. 2014. "Our Pets Are Family, Too." *Washingtion Times*. Accessed July 7, 2014. www.washingtontimes.com/news/2009/apr/21/our-pets-are-family-too/.
Davis, Bret W. 2007. *Heidegger and the Will: On the Way to Gelassenheit*. Evanston, IL: Northwestern University Press.
Dedios, John. 2011a. "A West World of Outreach, Biosphere 2 Is a Zion for Science." *Tucson Weekly*, October 5. www.tucsonweekly.com/TheRange/archives/2011/10/04/a-west-world-of-outreach-biosphere-2-a-zion-for-science. Last Accessed July 12, 2013.
———. 2011b. "Biosphere 2: A Model City for Sustainability (Audio included)." *Tucson Weekly*, November 13. www.tucsonweekly.com/TheRange/archives/2011/11/10/biosphere-2-a-model-city-for-sustainability-audio-and-slideshow-included. Last Accessed July 12, 2013.
Deleuze, Gilles. 2004. "Desert Islands." In *Desert Islands: And Other Texts, 1953–1974*, trans. Mike Taormina, 9–14. Los Angeles: Semiotext(e).
Denayer, W. "World and Mind in Hannah Arendt's Political Phenomenology." *Acta Politica* 29 (1): 37–49.
Derrida, Jacques. 1977. *Limited Inc*. Evanston, IL: Northwestern University Press.
———. 1981. "Economimesis." Trans. R. Klein. *Diacritics* 11, no. 2 (Summer): 3–25.
———. 1984. "No Apocalypse, Not Now (Full Speed Ahead, Seven Missiles, Seven Missives." Trans. Catherine Porter and Philip Lewis. *Diacritics* 14 (2): 20–31.

———. 1987. "Geschlecht II: Heidegger's Hand." In *Deconstruction and Philosophy: The Texts of Jacques Derrida*, ed. John Sallis, trans. John P. Leavey Jr., 161–196. Chicago: University of Chicago Press.

———. 1989. *Of Spirit: Heidegger and the Question*. Trans. Geoffrey Bennington and Rachel Bowlby. Chicago: University of Chicago Press.

———. 1991. "Geschlecht: Sexual Difference, Ontological Difference." In *A Derrida Reader*, ed. Peggy Kamuf, pp. 380–402. Trans. Ruben Bevezdivin. New York: Columbia University Press.

———. 1993. "Heidegger's Ear: Philopolemology (Geschlecht IV)." In *Reading Heidegger: Commemorations,* ed. John Sallis, 163–218. Bloomington: Indiana University Press.

———. 1998. "Geopsychoanalysis . . . 'and the Rest of the World.'" Trans. Donald Nicholson-Smith. In *The Psychoanalysis of Race*, ed. Christopher Lane, 65–90. New York: Columbia University Press.

———. 2000. *Of Hospitality*. Trans. Rachel Bowlby. Stanford: Stanford University Press.

———. 2001. *On Cosmopolitanism and Forgiveness*. Trans. Mark Dooley and Michael Hughes. New York: Routledge.

———. 2002a. "The Future of the Profession or the University Without Condition." In *Jacques Derrida and the Humanities*, ed. Tom Cohen, 24–57. New York: Cambridge University Press.

———. 2002b. "Globalization, Peace, and Cosmopolitanism." In *Negotiations: Interventions and Interviews, 1971–2000*, ed. Elizabeth Rottenberg, 371–386. Stanford: Stanford University Press.

———. 2003a. "Autoimmunity: Real and Symbolic Suicides." In *Philosophy in a Time of Terror: Dialogues with Jürgen Habermas and Jacques Derrida*, ed. Giovanna Borradon, 85–136. Chicago: University of Chicago Press.

———. 2003b. *Chaque fois unique, la fin du monde*. Paris: Galilee.

———. 2003c. "The 'World' of the Enlightenment to Come." *Research in Phenomenology* 33 (1): 30.

———. 2005a. "What Does It Mean to Be a French Philosopher Today?" In *Paper Machine*, trans. Rachel Bowlby, 112–120. Stanford: Stanford University Press.

———. 2005b. *Sovereignties in Question: The Poetics of Paul Celan*. New York: Fordham University Press.

———. 2005c. *Rogues: Two Essays on Reason*. Trans. Pascale-Anne Brault and Michael Naas. Stanford, Calif.: Stanford University Press.

———. 2008. *The Animal That Therefore I Am*. Trans. David Wills. New York: Fordham University Press.

———. 2009. *The Beast and the Sovereign, Volume I*. Trans. Geoffrey Bennington. Chicago: University of Chicago Press.

———. 2010. *Learning to Live Finally: The Last Interview*. Trans. Jean Birnbaum. Brooklyn, NY: Melville House.

———. 2011. *The Beast and the Sovereign, Volume II*. Trans. Geoffrey Bennington. Chicago: University of Chicago Press.

———. 2012. *Specters of Marx*. New York: Routledge.
Dick, Steven J. 1982. *Plurality of Worlds. The Origins of the Extra-Terrestrial Life Debate from Democritus to Kant*. Cambridge: Cambridge University Press.
Dickey, James. 1969. "The Moon Ground." *Life* (August 11): 16c.
Dickinson, Colby. 2011. "The Logic of the "As If" and the (Non)Existence of God: An Inquiry Into the Nature of Belief in the Work of Jacques Derrida." *Derrida Today* 4 (1): 86–106.
Dietz, Mary G. 2010. "Feminist Receptions of Hannah Arendt." In *Feminist Interpretations of Hannah Arendt*, ed. Bonnie Honig, 17–50. State College: Penn State Press.
Disch, Lisa. 1994. *Hannah Arendt and the Limits of Philosophy*. Ithaca: Cornell University Press.
———. 1997. "Please Sit Down, but Don't Make Yourself at Home: Arendtian "Visiting" and the Prefigurative Politics of Consciousness-raising." In *Hannah Arendt and the Meaning of Politics*, ed. Craig Calhoun and John McGowan, 132–164. Minneapolis: University of Minnesota Press.
Dobbs, Favid. 2013. "Zeal for Play May Have Propelled Human Evolution." *New York Times*. April 22, http://www.nytimes.com/2013/04/23/science/zeal-for-play-may-have-propelled-human-evolution.html?pagewanted=all
Doyle, Arthur Conan. 1990. *When the World Screamed and Other Stories*. San Francisco: Chronicle.
Dreyfus, Hubert. 1975. "The Priority of *the* World to *My* World: Heidegger's Answer to Husserl (and Sartre). *Man and World* 8 (2): 121–130.
———. 2003. "Further Reflections on Heidegger, Technology, and the Everyday." *Bulletin of Science Technology Society* 23 (5): 339–349.
———. 2007. "Heidegger's Ontology of Art." In *A Companion to Heidegger*, ed. Hubert Dreyfus and Mark A. Wrathall, 407–419. Oxford: Blackwell.
Duits, Rufus A. 2009. *Raising the Question of Being: A Unification and Critique of the Philosophy of Martin Heidegger*. Boca Raton: Universal.
Durst, Margarete. 2004. "Birth and Natality in Hannah Arendt." *Analecta Husserliana* 70: 777–797.
Dutoit, Thomas. 2012. "Kant's Retreat, Hugo's Advance, Freud's Erection; or, Derrida's Displacements in His Death Penalty Lectures." *Southern Journal of Philosophy* 50:107–135.
Edwards, Jeffrey. 2000. "Self-Love, Anthropology, and Universal Benevolence in Kant's Metaphysics of Morals." *Review of Metaphysics* 53 (June): 887–914.
———. 2011. "'The Unity of All Places on the Face of the Earth': Original Community, Acquisition, and Universal Will in Kant's Doctrine of Right." In *Reading Kant's Geography*, translated by Stuart Elden and Eduardo Mendieta, 233–64. Albany: SUNY Press.
Emad, Parvis. 1985. "Boredom as Limit and Disposition." *Heidegger Studies* 1:63–78.
Feil, Ken. 2005. *Dying for a Laugh: Disaster Movies and the Camp Imagination*. Wesleyan: Wesleyan University Press.
Felman, Shoshana, and Dori Laub. 1992. *Testimony: Crises of Witnessing in Literature, Psychoanalysis and History*. New York: Routledge.

Fenves, Peter. 2003. *Late Kant: Towards Another Law of the Earth*. New York: Routledge.
Fioretos, Aris. 1990. "Nothing: Reading Paul Celan's 'Engfuhrung.'" *Comparative Literature Studies* 27 (2): 158–168.
Firat, A. F., and Alladi Venkatesh. 1995. "Liberatory Postmodernism and the Reenchantment of Consumption." *Journal of Consumer Research* 22 (3): 239–267.
Føllesdal, Dagfinn. 1978. "Husserl and Heidegger on the Role of Actions in the Constitution of the World." *Essays in Honor of Jaako Hintikka*, ed. Esa Saarinen, Risto Hilpinen, Ilkka Niinuluoto, Merill Provence Hintikka, 365–378. Hingham, MA: Springer Netherlands.
Foltz, Bruce V. 1995. *Inhabiting the Earth: Heidegger, Environmental Ethics, and the Metaphysics of Nature*. Atlantic Highlands, NJ: Humanities.
Fóti, Véronique M. 1998. "Heidegger and 'the Way of Art': The Empty Origin and Contemporary Abstraction." *Continental Philosophy Review* 31 (4): 337–351.
Franck, Didier. 1991. "Being and the Living." In *Who Comes After the Subject?* ed. Eduardo Cadava, Peter Connor, and Jean-Luc Nancy, 135–147. New York: Routledge.
Fried, Gregory. 2000. *Heidegger's Polemos: From Being to Politics*. New Haven: Yale University Press.
Fritsch, Matthias. 2011. "Taking Turns: Democracy to Come and Intergenerational Justice." *Derrida Today* 4 (2): 148–172.
Fujita, Hisashi. 2012. "University with Conditions: A Deconstructive Reading of Derrida's 'the University Without Condition.'" *Southern Journal of Philosophy* 50 (2): 250–272.
Gadamer, Hans-Georg. 1983. "What Is Practice? The Conditions of Social Reason." In *Reason in the Age of Science*, trans. Frederick G. Lawrence, 69–87. Cambridge: MIT Press.
———.1994. *Heidegger's Ways*. Trans. John W. Stanley. Albany: SUNY.
———. 1997. "Who Am I and Who Are You?" In *Gadamer on Celan: "Who Am I and Who Are You?" and Other Essays*, ed. and trans. Richard Heinemann and Bruce Krajewski, 67–126. Albany: SUNY Press.
Gall, Robert S. 2003. "Interrupting Speculation: The Thinking of Heidegger and Greek Tragedy." *Continental Philosophy Review* 36 (2): 177–194.
Garb, Y. J. 1985. "The Use and Misuse of the Whole Earth Image." *Whole Earth Review* (March): 18–25.
Garrard, Greg. 2012. "Worlds Without Us: Some Types of Disanthropy." *Substance* 41 (1): 40–60.
Gauthier, David J. 2011. *Martin Heidegger, Emmanuel Levinas, and the Politics of Dwelling*. Lanham, MD: Lexington.
George, Theodore. 2012. "Thing, Object, Life." *Research in Phenomenology* 42 (1): 18–34.
Ghosh, Ranjan. 2012. "Globing the Earth: The New Eco-logics of Nature." *Substance* 41 (1): 3–14.
Gibbs, Paul. 2011. "The Concept of Profound Boredom: Learning from Moments of Vision." *Studies in Philosophy and Education* 30 (6): 601–613.

Glazebrook, Trish. 2000. *Heidegger's Philosophy of Science*. New York: Fordham University Press.
Goodrich, Peter, ed. 2008. *Derrida and Legal Philosophy*. Basingstoke: Palgrave Macmillan.
Gordon, Hayim and Shlomit Tamari. 2004. *Maurice Merleau-Ponty's Phenomenology of Perception: A Basis for Sharing the Earth*. Westport, CT: Praeger.
Gordon, Rivca. 2003. "Questioning Martin Heidegger's Thinking on Boredom." *Philosophical Inquiry* 25 (1–2): 125–134.
Gosetti-Ferencei, Jennifer Anna. 2002. "Tragedy and Truth in Heidegger and Jaspers." *International Philosophical Quarterly* 42 (3): 301–314.
———. 2004. *Heidegger, Hölderlin, and the Subject of Poetic Language: Toward a New Poetics of Dasein*. New York: Fordham University Press.
———. 2012. "The World and Image of Poetic Language: Heidegger and Blanchot." *Continental Philosophy Review* 45 (2): 189–212.
Gottlieb, Susannah Young-Ah. 2003. *Regions of Sorry: Anxiety and Messianism in Hannah Arendt and W. H. Auden*. Palo Alto: Stanford University Press.
Griffiths, Dominic. 2007. "Reading Elements of the Later Heidegger as Myth." *Phronimon* 8 (2): 25–34.
———. 2012. "'Now and in England': Four Quartets, Place, and Martin Heidegger's Concept of Dwelling." *Yeats Eliot Review* 29 (1/2): 3–18.
Griffiths, Dominic and Maria Prozesky. 2010. "The Politics of Dwelling: Being White/Being South African." *Africa Today* 56 (4): 22–41.
Grumley, John. 1996. "'Worldliness' in the Modern World: Heller and Arendt." *Thesis Eleven* 47 (November): 73–88.
Guenther, Lisa. 2006. "The Body Polity: Arendt on Time, Natality, and Reproduction." In *The Gift of the Other: Levinas and the Politics of Reproduction*, 29–48. Albany, N.Y.: State University of New York Press.
Haar, Michel. 1993. *The Song of the Earth: Heidegger and the Grounds of the History of Being*. Bloomington: Indiana University Press.
Hall, Harrison. 1993. "Intentionality and World: Division I of *Being and Time*." In *The Cambridge Companion to Heidegger*, ed. Charles B. Guignon, 122–140. Cambridge: Cambridge University Press.
Hammer, Espen. 2004. "Being Bored: Heidegger on Patience and Melancholy." *British Journal for the History of Philosophy* 12 (2): 277–295.
———. 2008. "Heidegger's Theory of Boredom." *Graduate Faculty Philosophy Journal* 29 (1): 199–225.
Hammermeister, Kai. 2000. "Heimat in Heidegger and Gadamer." *Philosophy and Literature* 24 (2): 312–326.
Hanlon, Michael. 2009. "Is This Haunting Picture Proof That Chimps Really DO Grieve?" *Daily Mail Online*, October 27. www.dailymail.co.uk/sciencetech/article-1223227/Is-haunting-picture-proof-chimps-really-DO-grieve.html.
Hansen, Mark. 2000. *Embodying Technesis: Technology Beyond Writing*. Ann Arbor: University of Michigan Press.

Harafin, Patrick. 2013. "'As Nobody I Was Sovereign': Reading Derrida Reading Blanchot." *Societies* 3:43–51.

Harman, Graham. 2007. *Heidegger Explained: From Phenomenon to Thing.* Chicago: Open Court.

———. 2009. "Dwelling with the Fourfold." *Space and Culture* 12 (3): 292–302.

Harries, Karsten. 2009. *Art Matters: A Critical Commentary on Heidegger's "The Origin of the Work of Art.* Dordrecht: Springer.

The Harris Poll. 2007. "Pets and 'Members of the Family' and Two-Thirds of Pet Owners Buy Their Pets Holiday Presents." No. 120.

Harrison, Paul. 2007. "The Space Between Us: Opening Remarks on the Concept of Dwelling." *Environment and Planning D: Society and Space* 25 (4): 625–647.

Hayes, Josh. 2007. "Deconstructing Dasein: Heidegger's Earliest Interpretations of Aristotle's de Anima." *Review of Metaphysics* 61 (2): 263–293.

Hedrick, Todd. 2008. "Race, Difference, and Anthropology in Kant's Cosmopolitanism." *Journal of the History of Philosophy* 46 (2): 245–268.

Heidegger, Martin. 1968 [1951–1952]. *What Is Called Thinking?* Trans. Fred D. Wieck and J. Glenn Gray. New York: Harper and Row.

———. 1969 [1955–1957]. *Identity and Difference.* Trans. Joan Stambaugh. New York: Harper and Row.

———. 1971a [1951]. "Building Dwelling Thinking." In *Poetry, Language, Thought,* trans. Albert Hofstadter. New York: Harper and Row.

———. 1971b [1950–1959]. *On the Way to Language.* Trans. Peter D. Hertz. New York: Harper and Row.

———. 1971c [1935–1936]. "The Origin of the Work of Art." In *Poetry, Language, Thought,* trans. Albert Hofstadter. New York: Harper and Row.

———. 1971d [1951]. "Poetically Man Dwells . . . " In *Poetry, Language, Thought,* trans. Albert Hofstadter. New York: Harper and Row.

———. 1971e [1951]. "The Thing." In *Poetry, Language, Thought,* trans. Albert Hofstadter. New York: Harper and Row.

———. 1972 [1962–1964]. *On Time and Being.* Trans. Joan Stambaugh. New York: Harper and Row.

———. 1973a [1969]. "Art and Space." Trans. Charles H. Seibert. *Man and World* 6 (1): 3–8.

———. 1973b. "Overcoming Metaphysics." In *The End of Philosophy,* trans. Joan Stambaugh, 84–111. New York: Harper and Row.

———. 1976. *The Piety of Thinking: Essays by Martin Heidegger.* Trans. James Hart and John Maraldo. Bloomington: Indiana University Press.

———. 1977a [1938]. "The Age of the World Picture." In *The Question Concerning Technology and Other Essays,* trans. William Lovitt, 115–154. New York: Harper and Row.

———. 1977b [1954]. "The Question Concerning Technology." In *The Question Concerning Technology and Other Essays,* trans. William Lovitt, 3–35. New York: Harper and Row.

———. 1981 [1966]. "Only a God Can Save Us: The Spiegel Interview." In *Heidegger: The Man and the Thinker*, ed. Thomas Sheehan, 4567. Chicago: Precedent.

———. 1983 [1957]. "Hebel–Friend of the House." Trans. B. V. Foltz and M. Heim. *Contemporary German Philosophy* 3:89–101.

———. 1985. "The Self-Assertion of the German University: Address, Delivered on the Solemn Assumption of the Rectorate of the University Freiburg, the Rectorate 1933/34: Facts and Thoughts." Trans. Karsten Harries. *Review of Metaphysics* 38 (3): 467–502.

———. 1992. *Parmenides*. Trans. Andre Schuwer and Richard Rojcewicz. Bloomington: Indiana University Press.

———. 1993 [1946]. "Letter on Humanism." In *Basic Writings*, ed. David Krell, 213–266. New York: HarperCollins.

———. 1995 [1929–1930]. *The Fundamental Concepts of Metaphysics: World, Finitude, Solitude*. Trans. William McNeill and Nicholas Walker. Bloomington: Indiana University Press.

———. 1996a [1927]. *Being and Time*. Trans. Joan Stambaugh. Albany: State University of New York Press.

———. 1996b [1942]. *Hölderlin's Hymn "The Ister."* Trans. William McNeill and Julia Davis. Bloomington: Indiana University Press.

———. 1997. *Phenomenological Interpretation of Kant's Critique of Pure Reason*. Trans. Parvis Emad and Kenneth Maly. Bloomington: Indiana University Press.

———. 1998a [1949]. "Introduction to "What Is Metaphysics?" In *Pathmarks*, ed. William McNeill. Cambridge: Cambridge University Press.

———. 1998b [1928]. "On the Essence of Ground." In *Pathmarks*, ed. William McNeill. Cambridge: Cambridge University Press.

———. 1998c [1930]. "On the Essence of Truth." In *Pathmarks*, ed. William McNeill. Cambridge: Cambridge University Press.

———. 1998d [1943]. "Postscript to 'What Is Metaphysics?'" In *Pathmarks*, ed. William McNeill. Cambridge: Cambridge University Press.

———. 2000a. "As When on Holiday." In *Elucidations of Hölderlin's Poetry*, trans. Keith Hoeller. New York: Humanity.

———. 2000b. "Hölderlin and the Essence of Poetry." In *Elucidations of Hölderlin's Poetry*, trans. Keith Hoeller. New York: Humanity.

———. 2000c [1935]. *An Introduction to Metaphysics*. Trans. Gregory Fried and Richard Polt. New Haven: Yale University Press.

———. 2001 [1971]. "What Are Poets For?" In *Poetry, Language, Thought*, trans. Albert Hofstadter, 209–227. New York: HarperCollins.

———. 2005. *Sojourn*. Trans. John Panteleimon Manoussakis. Albany: SUNY Press.

———. 2008. "The Idea of Philosophy and the Problem of Worldview." In *Towards the Definition of Philosophy*, trans. Ted Sadler, 1–99. New York: Continuum.

———. 2009. *Ontology: The Hermeneutics of Facticity*. Trans. John Van Buren. Bloomington: Indiana University Press.

———. 2012a [1949]. *Bremen and Freiburg Lectures: Insight Into That Which Is and Basic Principles of Thinking*. Trans. Andrew J. Mitchell. Bloomington: Indiana University Press.

———. 2012b [1936–1938]. *Contributions to Philosophy (of the Event)*. Trans. Richard Rojcewicz and Daniela Vallega-Neu. Bloomington: Indiana University Press.

Heise, Ursula K. 2012. "Journeys Through the Offset World: Global Travel Narratives and Environmental Crisis." *Substance* 41 (1): 61–76.

Held, Klaus. 1992. "The Finitude of the World: Phenomenology in Transition from Husserl to Heidegger." Trans. Anthony J. Steinbock. In *Ethics and Danger: Essays on Heidegger and Continental Thought*, ed. Arleen B. Dallery, Charles E. Scott et al., 187–198. Albany: SUNY Press.

———. 1998. "Sky and Earth as Invariants of the Natural Life-World." In *Phenomenology of Interculturality and Life-World*, ed. Ernst Wolfgang Orth and Chan-Fei Cheung, 21–41. Freiburg: Karl Alber.

Highfield, Roger. 2006. "Elephants Show Compassion in Face of Death." *Telegraph*, August 14.

Himanka, Juha. 2000. "Does the Earth Move?" *Philosophical Forum* 31:57–83.

———. 2005. "Husserl's Argumentation for the Pre-Copernican View of the Earth." *Review of Metaphysics* 58 (3): 621–644.

Honig, Bonnie. 1991. "Declarations of Independence: Arendt and Derrida on the Problem of Founding a Republic." *American Political Science Review* 85 (1) (March): 97.

———. 2006. "Another Cosmopolitanism? Law and Politics in the New Europe." In *Another Cosmopolitanism*, ed. Robert Post. Oxford: Oxford University Press.

Hoffman, Piotr. 1993. "Death, Time, History: Division II of *Being and Time*." In *The Cambridge Companion to Heidegger*, ed. Charles B. Guignon, 195–214. Cambridge: Cambridge University Press.

Huhn, Tom. 1996. "The Movement of Mimesis: Heidegger's 'Origin of the Work of Art' in Relation to Adorno and Lyotard." *Philosophy and Social Criticism* 22 (4): 45–69.

Huntington, Patricia J. 1998. "Heidegger's Apolitical Nostalgia for Immediacy." In *Ecstatic Subjects, Utopia, and Recognition: Kristeva, Heidegger, Irigaray*, ed. Patricia J. Huntington, 205–232. Albany: SUNY Press.

Husserl, Edmund. 1981a. "Foundational Investigations of the Phenomenological Origin of the Spatiality of Nature." Trans. Frederick Kersten. In *Husserl: Shorter Works*, ed. Peter McCormick and Frederick A. Elliston, 213–221. Notre Dame, IN: University of Notre Dame Press.

———. 1981b. "The World of the Living Present." In *Husserl: The Shorter Works*, ed. Peter McCormick and Frederick Elliston, 238–250. Notre Dame, IN: University of Notre Dame Press.

———. 1989. *La terre ne se meut pas*. Trans. Didier Frank, Jean-François Lavigne, and Dominique Pradelle. Paris: Minuit.

———. 2003 [1934]. "Foundational Investigations of the Phenomenological Origin of the Spatiality of Nature: The Originary Ark, the Earth, Does Not Move." In Maurice Merleau-Ponty, *Husserl at the Limits of Phenomenology*, 117–131. Evanston: Northwestern University Press.

Hutchings, Patrick. 2012. "'The Origin of the Work of Art': Heidegger." *Sophia* 51 (4): 465–478.

Hribal, Jason. 2011. *Fear of the Animal Planet: The Hidden History of Animal Resistance*. Oakland, CA: AK.

Ihde, Don. 1990. "Husserl's Galileo." In *Technology and the Lifeworld: From Garden to Earth*, 34–37. Bloomington: Indiana University Press.

———. 1997. "Whole Earth Measurements: How Many Phenomenologists Does It Take to Detect a 'Greenhouse Effect'?" *Philosophy Today* 41 (1): 128–134.

Iliea, Laura T. 2007. "Hermeneutics of the Artwork and the End of the Age of Nostalgia: From Oblivion of Being (Heidegger) to Its Enhancement (Gadamer)." *RES: Anthropology and Aesthetics* 51 (Spring): 216–225.

Ingold, Tim. 2002. "Globes and Spheres." In *The Perception of the Environment: Essays on Livelihood, Dwelling, and Skill*, 209–218. London: Routledge.

———. 2007. "Earth, Sky, Wind, and Weather." *Journal of the Royal Anthropological Institute* 13:S19–S38.

Introna, Lucas D. and Fernando M. Ilharco. 2005. "Phenomenology, Screens, and the World: A Journey with Husserl and Heidegger Into Phenomenology." In *Social Theory and Philosophy for Information Systems*, ed. John Mingers and Leslie P. Willcocks, 56–102. Wiley.

Irigaray, Luce. 1985. *Speculum of the Other Woman*. Trans. Gillian C. Gill. Ithaca: Cornell University Press.

———. 2008. *Sharing the World*. London: Continuum.

Irwin, Ruth. 2008. *Heidegger, Politics, and Climate Change: Risking It All*. London: Continuum.

———. 2010a. "Climate Change and Heidegger's Philosophy of Science." *Essays in Philosophy* 11 (1): 16–430.

———. 2010b. "Reflections on Modern Climate Change and Finitude." In *Climate Change and Philosophy: Transformational Possibilities*, ed. Ruth Irwin, 48–72. London: Continuum.

Isaac, Jeffrey. 2002. "Hannah Arendt on Human Rights and the Limits of Exposure, or Why Noam Chomsky Is Wrong about the Meaning of Kosovo." *Social Research* 69 (2): 505–537.

Jaeger, Hans. 1958. "Heidegger and the Work of Art." *Journal of Aesthetics and Art Criticism* 17 (1): 58–71.

James, Simon P. 2007. "Merleau-Ponty, Metaphysical Realism, and the Natural World." *International Journal of Philosophical Studies* 15 (4): 501–519.

Janover, Michael. 2005. "The Limits of Forgiveness and the Ends of Politics." *Journal of Intercultural Studies* 26 (3): 221.

Jaran, François. 2010. "Toward a Metaphysical Freedom: Heidegger's Project of a Metaphysics of Dasein." *International Journal of Philosophical Studies* 18 (2): 205–227.

Jasanoff, Sheila. 2004. "Heaven and Earth: The Politics of Environmental Images." In *Earthly Politics: Local and Global in Environmental Governance*, ed. Sheila Jasanoff and Marybeth Long Martello, 31–52. Cambridge: MIT Press.

Jennings, Theodore W. 2006. *Reading Derrida/Thinking Paul: On Justice*. Stanford: Stanford University Press.

Johnson, Galen A. 1989. *Earth and Sky, History and Philosophy: Island Images Inspired by Husserl and Merleau-Ponty*. New York: Peter Lang.

Joronen, Mikko. 2008. "The Technological Metaphysics of Planetary Space: Being in the Age of Globalization." *Environment and Planning D: Society and Space* 26 (4): 596–610.

———. 2011. "Dwelling in the Sites of Finitude: Resisting the Violence of the Metaphysical Globe." *Antipode* 43 (4): 1127–1154.

Kain, Patrick. 2009. "Kant's Defense of Human Moral Status." *Journal of the History of Philosophy* 47 (1): 59–101.

Kakoudaki, Despina. 2002. "Spectacles of History: Race Relations, Melodrama, and the Science Fiction/Disaster Film." *Camera Obscura* 50, 17 (2): 109–153.

Kateb, George. 1984. *Hannah Arendt: Politics, Conscience, Evil*. Totowa, NJ: Rowman and Allanheld.

Kant, Immanuel. 1967. *Philosophical Correspondence, 1759–1799*. Ed. and trans. Arnulf Zweig. Chicago: University of Chicago Press.

———. 1996a [1797]. *The Metaphysics of Morals*. Ed. and trans. Mary J. Gregor. Cambridge: Cambridge University Press.

———. 1996b [1793]. "On the Common Saying: That May Be Correct in Theory, but It Is of No Use in Practice." In *Practical Philosophy*, ed. and trans. Mary J. Gregor, 273–310. Cambridge: Cambridge University Press.

———. 1996c [1795]. "Perpetual Peace." In *Practical Philosophy*, ed. and trans. Mary J. Gregor, 311–352. Cambridge: Cambridge University Press.

———. 2001. *Prolegomena to Any Future Metaphysics*. 2d ed. eHackett.

———. 2003. *Political Writings*. Ed. H. S. Reiss. Cambridge: Cambridge University Press.

———. 2006. *Anthropology from a Pragmatic Point of View*. Ed. Robert B. Louden. Cambridge: Cambridge University Press.

———. 2007a. *Anthropology, History, and Education*. Ed. Günter Zöller and Robert B. Louden, trans. Mary Gregor et al. Cambridge: Cambridge University Press.

———. 2007b [1784]. "Idea for a Universal History with a Cosmopolitan Aim." In *Anthropology, History, and Education*, ed. Günter Zöller and Robert B. Louden; trans. Allen W. Wood, 107–120. Cambridge: Cambridge University Press.

———. 2011 [1760]. "Thoughts on the Occasion of Mr. Johann Friedrich von Funk's Untimely Death." In *Immanuel Kant: Observations on the Feeling of the Beautiful and Sublime and Other Writings*, ed. Patrick Frierson and Paul Guyer, 3–10. Cambridge: Cambridge University Press.

———. 2012a. *Natural Science*. Ed. Eric Watkins. Cambridge University Press.

———. 2012b [1754]. "The Question, Whether the Earth Is Ageing, Considered from a Physical Point of View." Trans. Olaf Reinhardt. In *Natural Science*, ed. Eric Watkins, 165–181. Cambridge: Cambridge University Press.

———. 2012c [1755]. "Universal Natural Theory and Theory of the Heavens." Trans. Olaf Reinhardt. In *Natural Science*, ed. Eric Watkins, 182–308. Cambridge: Cambridge University Press.

Karnoouh, Claude. 2002. "On the Genealogy of Globalization." *Telos* 124:183–192.

Kellert, Stephen R., and Edward O. Wilson, eds. 1993. *The Biophilia Hypothesis*. Washington, DC: Island.

Kateb, George. 1977. "Freedom and Worldliness in the Thought of Hannah Arendt." *Political Theory* 5 (2): 141–182.

Keane, Stephen. 2006. *Disaster Movies: The Cinema of Catastrophe*. London: Wallflower.

Kevles, Daniel J. 2013. "Can They Patent Your Genes?" *New York Review of Books*, March 7. www.nybooks.com/articles/archives/2013/mar/07/can-they-patent-your-genes/.

King, Geoff. "Just Like a Movie? 9/11 and Hollywood Spectacle." In *The Spectacle of the Real: From Hollywood to Reality TV and Beyond*, ed. Geoff King, 47–58. Bristol: Intellect.

Kirby, Rob. 2012. "Wild Elephants Gather Inexplicably, Mourn Death of 'Elephant Whisperer.'" *Delight Makers*. http://delightmakers.com/news-bleat/wild-elephants-gather-inexplicably-mourn-death-of-elephant-whisperer/.

Klaver, Irene J. 2012. "Authentic Landscapes at Large: Dutch Globalization and Environmental Imagination." *Substance* 41 (1): 92–108.

Klein, Julie. 2003. "Nature's Metabolism: On Eating in Derrida, Agamben, and Spinoza." *Research in Phenomenology* 33 (1): 186–217.

Kleingeld, Pauline. 1998. "Kant's Cosmopolitan Law: World Citizenship for a Global Order." *Kantian Review* 2:72–90.

———. 2003. "Kant's Cosmopolitan Patriotism." *Kant-Studien* 94 (3): 299–316.

Kockelmans, Joseph J. 1989. *Heidegger's "Being and Time": The Analytic of Dasein as Fundamental Ontology*. Lanham, MD: University Press of America.

Kreeft, Peter. 1971. "Zen in Heidegger's Gelassenheit." *International Philosophical Quarterly* 11 (4): 521–545.

Krell, David Farrell. 1992. "*Das Unheimliche*: Architectural Sections of Heidegger and Freud." *Research in Phenomenology* 22 (1): 43–61.

———. 2008. "Introduction to Building, Dwelling, Thinking." In *Basic Writings*, ed. David Krell, 343–346. New York: HarperCollins.

———. 2013. *Derrida and Our Animal Others: Derrida's Final Seminar, the Beast and the Sovereign*. Bloomington: Indiana University Press.

Kristeva, Julia. 2001. *Hannah Arendt*. Trans. Ross Guberman. Toronto: University of Toronto Press.

Kuperus, Gerard. 2007. "Attunement, Deprivation, and Drive: Heidegger and Animality." In *Phenomenology and the Non-Human Animal*, ed. Christian Lotz and Corrine Painter. Dordrecht, Netherlands: Springer.

Lacan, Jacques. 1981. *The Seminar of Jacques Lacan: Book XI, the Four Fundamental Concepts of Psychoanalysis*. Trans. Alan Sheridan. New York: Norton.

———. 1987a. "Introduction to the Names-of-the-Father Seminar." Trans. Jeffrey Mehlman. *October* 40 (Spring): 81–95.

———. 1987b. "Television." Trans. Denis Hollier, Rosalind Krauss, and Annette Michelson. *October* 40 (Spring): 6–50.

———. 2001. *Écrits: A Selection*. New York: Routledge.

———. 2004. "The Subversion of the Subject and the Dialectic of Desire in the Freudian Unconscious." In *Hegel and Contemporary Continental Philosophy*, ed. Denis King Keenan, 205–250. Albany: SUNY Press.

La Caze, Marguerite. 2007. "At the Intersection: Kant, Derrida, and the Relation Between Ethics and Politics." *Political Theory* 35 (6): 781–805.

Lacoue-Labarthe, Philippe. 1990. *Heidegger, Art, and Politics: The Fiction of the Political*. Oxford: Blackwell.

Lafont, Cristina. 2000. *Heidegger, Language, and World-Disclosure*. Cambridge: Cambridge University Press.

Lambert, Gregg. 2005. "What the Earth Thinks." In *Deleuze and Space*, ed. Ian Buchanan and Gregg Lambert, 220–239. Edinburgh: Edinburgh University Press.

Lang, Michael. 2006. "Globalization and Its History." *Journal of Modern History* 78 (4): 899–931.

Lash, Scott. 1998. "Being After Time: Towards a Politics of Melancholy." *Cultural Values* 2 (2–3): 305–319.

Lawlor, Leonard. 2006. *The Implications of Immanence: Toward a New Concept of Life*. New York: Fordham University Press.

———. 2007. *This Is Not Sufficient: An Essay on Animality and Human Nature in Derrida*. New York: Columbia University Press.

Lazier, Benjamin. 2011. "Earthrise; or, the Globalization of the World Picture." *American Historical Review* 116 (3): 602–630.

Lee, Kwang-Sae. 2012. "Heidegger's Seyn, Ereignis, and Dingen as Viewed from an Eastern Perspective." *Journal of Philosophical Research* 37:343–351.

Lévinas, Emmanuel. 1978. "Being and the Other: On Paul Celan." Trans. Stephen Melville. *Chicago Review* 29 (3): 16–22.

———. 1986. "Sur la mort chez Ernst Bloch." In *De Dieu qui vient a l'idee*. Paris: Vrin.

———. 1991. "Mourir pour." In *Entre nous*. Paris: Gassett.

———. 2000. *God, Death, and Time*. Trans. Bettina Bergo. Stanford: Stanford University Press.

Li, Victor. 2007. "Elliptical Interruptions: Or, Why Derrida Prefers Mondialisation to Globalization." *CR: The New Centennial Review* 7 (2): 141–154.

Linklater, Andrew. 1998. "Cosmopolitan Citizenship." *Citizenship Studies* 2 (1): 23–41.

Llewelyn, John. 1991. "Am I Obsessed by Bobby? (Humanism and the Other Animal)." In *Re-Reading Levinas*, ed. Robert Bernasconi and Simon Critchley, 234–245. Bloomington: Indiana University Press.

———. 1995. *Emmanuel Levinas: The Genealogy of Ethics*. New York: Routledge.

———. 2002. *Appositions of Jacques Derrida and Emmanuel Levinas*. Bloomington: Indiana University Press.
Loscerbo, John. 1981a. "Remarks Concerning Some Earlier Texts." In *Being and Technology: A Study in the Philosophy of Martin Heidegger*, 101–114. Hingham, MA: Springer Netherlands.
———. 1981b. "The Platonic 'IΔÉA and 'IΔÉIN." In *Being and Technology: A Study in the Philosophy of Martin Heidegger*, 43–58. Hingham, MA: Springer Netherlands.
Losurdo, Domenico. 2001. *Heidegger and the Ideology of War: Community, Death, and the West*. New York: Humanity.
Louden, Robert B. 2008. "Anthropology from a Kantian Point of View: Toward a Cosmopolitan Conception of Human Nature." *Studies in History and Philosophy of Science* 39:515–522.
Lovelock, James. 1988. *The Ages of Gaia a Biography of Our Living Earth*. Oxford: Oxford University Press.
———. 2000. *Gaia: A New Look at Life on Earth*. Oxford: Oxford University Press.
Lulofs, J. G. 1962. "Labour and Consumption." *Mens en Maatschappij* 37 (2): 101–112.
Lyotard, Jean François. 1991. "Can Thought Go on Without a Body?" In *The Inhuman: Reflections on Time*, trans. Geoffrey Bennington and Rachel Bowlby, 8–23. Stanford: Stanford University Press.
Macauley, David. 1992. "Out of Place and Outer Space: Hannah Arendt on Earth Alienation: An Historical and Critical Perspective." *Capitalism Nature Socialism* 3 (4): 19–45.
———. 1996. "Hannah Arendt and the Politics of Place: From Earth Alienation to Oikos." In *Minding Nature: The Philosophers of Ecology*, ed. David Macauley, 102–133. New York: Guilford.
MacLeish, Archibald. 1944. "The Image of Victory." In *Compass of the World: A Symposium of Political Geography*, ed. H. W. Weigert and V. Stefansson, 1–11. London: George G. Harrap.
———. 1968. "Riders on Earth Together, Brothers in Eternal Cold." *New York Times*, December 25.
———. 1976. "Voyage to the Moon." In *New and Collected Poems, 1917–1976*, 17–18. New York: Knopf.
Malpas, Jeff. 2006. *Heidegger's Topology: Being, Place, World*. Cambridge: MIT Press.
———. 2011. "Philosophy's Nostalagia." In *Philosophy's Moods: The Affective Grounds of Thinking*, ed. Hagi Kenaan and Ilit Ferber, 87–101. Hingham, MA: Springer Netherlands.
———. 2012. *Heidegger and the Thinking of Place: Explorations in the Topology of Being*. Cambridge: MIT Press.
Mann, Bonnie. 2005. "World Alienation in Feminist Thought: The Sublime Epistemology of Emphatic Anti-Essentialism." *Ethics and the Environment* 10 (2): 45–74.
Mansikka, Jan-Erik. 2009. "Can Boredom Educate Us? Tracing a Mood in Heidegger's Fundamental Ontology from an Educational Point of View." *Studies in Philosophy and Education* 28 (3): 255–268.

Mardellat, Patrick. 2011. "L'économie comme dévoilement de la 'vie nue' selon Hannah Arendt." *Economies et Societes* 44 (July): 1179–1202.

Marder, Michael. 2007. "Heidegger's 'Phenomenology of Failure' in *Sein und Zeit*." *Philosophy Today* 51 (1): 69–78.

Marder, Elissa. 2010. "Ewa Ziarek's Virtually Impossible Ethics." In *Recenterings of Continental Philosophy*, ed. Cynthia Willett and Len Lawlor. Selected Studies in Phenomenology and Existential Philosophy, vol. 35. Chicago: DePaul University. *Philosophy Today* 54, Supplement 2010: 51–58.

Marion, Jean-Luc. 1987. "L'ego et le Dasein Heidegger et la 'destruction' de Descartes dans 'Sein und Zeit'." *Revue de Métaphysique et de Morale* 92 (1): 25–53.

"Mars Colony Project Aims to Preserve Life Forms on the Red Planet." 2013. *Zeenews.com* April 24. http://zeenews.india.com/news/space/mars-colony-project-aims-to-preserve-life-forms-on_844396.html. Last accessed July 12, 2013.

Martel, James. 2008. "Amo: Volo Ut Sis: Love, Willing and Arendt's Reluctant Embrace of Sovereignty." *Philosophy and Social Criticism* 34 (3): 287–313.

Marwah, Inder S. 2012. "Bridging Nature and Freedom? Kant, Culture, and Cultivation." *Social Theory and Practice* 38 (3): 385–406.

Matuštík, Martin Beck. 2007. "'More Than All the Others': Meditation on Responsibility." *Critical Horizons* 8 (1): 47–60.

McComb, K., C. Moss, S. M. Durant, L. Baker, and S. Sayialel. 2001. "Matriarchs as Repositories of Social Knowledge in African Elephants." *Science* 292 (April 20): 491–494.

McKenzie, Jonathan. 2008. "Governing Moods: Anxiety, Boredom, and the Ontological Overcoming of Politics in Heidegger." *Canadian Journal of Political Science* 41 (3): 569–585.

McLuhan, Marshall. 1974. "At the Moment of Sputnik the Planet Became a Global Theater in Which There Are No Spectators But Only Actors." *Journal of Communication* 24:48–58.

McManus, Denis. 2012. "Heidegger and the Supposition of a Single, Objective World." *European Journal of Philosophy.* DOI: 10.1111/j.1468-0378.2012.00529.x.

McMullin, Irene. 2009. "Sharing the 'Now': Heidegger and the Temporal Co-constitution of World." *Continental Philosophy Review* 42 (2): 201–220.

McNeill, William. 1999. *The Glance of the Eye: Heidegger, Aristotle, and the Ends of Theory*. Albany: State University of New York Press.

———. 2007. *The Time of Life: Heidegger and Ethos*. Albany: SUNY Press.

McVeigh, Karen. 2013. "Life on Mars to Become a Reality in 2023, Dutch Firm Claims." *Guardian*, April 21. www.guardian.co.uk/science/2013/apr/22/mars-one-project-astronauts. Last Accessed July 12, 2013.

McWhorter, Ladelle and Gail Stenstad, ed. 1992. *Heidegger and the Earth: Essays in Environmental Philosophy*. Toronto: Toronto University Press.

Megill, Allan. 1985. "Heidegger's Aestheticism." In *Prophets of Extremity: Nietzsche, Heidegger, Foucault, Derrida*, 142–180. Berkeley: University of California Press.

Menke, Christoph. 2007. "The 'Aporias of Human Rights' and the 'One Human Right': Regarding the Coherence of Hannah Arendt's Argument." *Social Research* 74 (3): 739–762.
Merleau-Ponty, Maurice. 1964a. "The Philosopher and His Shadow." In *Signs*, trans. Richard C. McCleary, 159–182. Evanston, IL: Northwestern University Press.
———. 1964b. *The Primacy of Perception: And Other Essays on Phenomenological Psychology, the Philosophy of Art, History, and Politics*. Ed. James M. Edie. Evanston, IL: Northwestern University Press.
———. 1964c. *Sense and Non-Sense*. Trans. Hubert L. Dreyfus and Patricia Allen Dreyfus. Evanston, IL: Northwestern University Press.
———. 1969. *Humanism and Terror: An Essay on the Communist Problem*. Trans. John O'Neill. Boston: Beacon.
———. 1973. *The Prose of the World*. Ed. Claude Lefort, trans. John O'Neill. Evanston, IL: Northwestern University Press.
———. 1979. *Consciousness and the Acquisition of Language*. Trans. Hugh J. Silverman. Evanston, IL: Northwestern University Press.
———. 1988. *In Praise of Philosophy and Other Essays*. Trans. John Wild and James M. Edie. Evanston, IL: Northwestern University Press.
———. 1992. "Husserl's Concept of Nature." In *Texts and Dialogues*. Ed. Hugh J. Silverman and James Barry Jr., trans. Drew Leder, 162–168. Atlantic Highlands, NJ: Humanities Press.
———. 2002. *Phenomenology of Perception*. Trans. Colin Smith. New York: Routledge.
———. 2003. *Nature: Course Notes from the Collège de France*. Trans. Robert Vallier. Evanston, IL: Northwestern University Press.
———. 2004. *The World of Perception*. Trans. Oliver Davis. New York: Routledge.
———. 2010. *Child Psychology and Pedagogy: The Sorbonne Lectures, 1949–1952*. Trans. Talia Welsh. Evanston, IL: Northwestern University Press.
———. 2010. *Institution and Passivity: Course Notes from the Collège de France (1954–1955)*. Trans. Leonard Lawlor and Heath Massey. Evanston, IL: Northwestern University Press.
Mickey, Sam. 2008. "Cosmological Postmodernism in Whitehead, Deleuze, and Derrida." *Process Studies* 37 (2): 24–44.
Miles, Margaret. 2002. "Volo Ut Sis: Arendt and Augustine." *Dialog* 41 (3): 221–30.
Miller, Elaine P. 2002. "Kant: The English Garden." In *The Vegetative Soul: From Philosophy of Nature to Subjectivity in the Feminine*, 19–44. Albany: SUNY Press.
———. 2005. "'The World Must Be Romanticised . . .': The (Environmental) Ethical Implications of Schelling's Organic Worldview." *Environmental Values* 14 (3): 295–316.
Miller, J. Hillis. 2012. "How to (Un)Globe the Earth in Four Easy Lessons." *Substance* 41 (1): 15–29.
Mills, Jon. 1997. "The False Dasein: From Heidegger to Sartre and Psychoanalysis." *Journal of Phenomenological Psychology* 28 (1): 42–65.

Misek, Richard. 2010. "Dead Time: Cinema, Heidegger, and Boredom." *Continuum: Journal of Media and Cultural Studies* 24 (5): 777–785.

Mitchell, Andrew J. 2005. "Heidegger and Terrorism." *Research in Phenomenology* 35 (1): 181–218.

———. 2009. "The Fourfold." In *Martin Heidegger: Key Concepts*, ed. Bret W. Davis, 208–218. Durham: Acumen.

———. 2011. "Towards a Heideggerian Floristics: Rethinking Organism in the Late Work." *Proceedings of the 45th Annual Meeting of the Heidegger Circle* 45:4–14.

———. 2012. "Translator's Foreword." In *Heidegger's Bremen and Freiburg Lectures*, vi–xvi. Bloomington: Indiana University Press.

———. 2014. *The Fourfold: Reading the Late Heidegger*. Evanston, IL: Northwestern University Press.

Mølbak, Rune L. 2011. "Lived Experience as a Strife Between Earth and World." *Journal of Theoretical and Philosophical Psychology* 31 (4): 207–222.

Morriston, Wesley. 1972. "Heidegger on the World." *Man and World* 5 (4): 452–467.

Mortari, Luigina. 1994. "Educating Ourselves to Think with a View to 'Ecologically Inhabiting the Earth' in Light of the Thought of Hannah Arendt." *Australian Journal of Environmental Education* 10:91–111.

Morton, Joseph. 2008. "The Semiotic Basis of Heidegger's *Dasein*." *Semiotics*, pp. 589–596.

Moskowitz, Clara. 2013. "Want to Live on Mars? Private Martian Colony Project Seeks Astronauts." *Space.com*, April 22. www.space.com/20758-private-mars-one-colony-astronauts.html. Last Accessed July 12, 2013.

Mugerauer, Robert. 2008. *Heidegger and Homecoming: The Leitmotif in the Later Writings*. Toronto: University of Toronto Press.

Murphy, Patrick D. 2012. "The Procession of Identity and Ecology in Contemporary Literature." *SubStance* 41 (1): 77–99.

Naas, Michael. 2005. "'Alors, Qui Etes-Vous?' Jacques Derrida and the Question of Hospitality." *SubStance* 34 (1): 6–17.

———. 2008. *Derrida from Now On*. New York: Fordham University Press.

———. 2012a. "If You Could Take Just Two Books . . . ": Jacques Derrida at the Ends of the World with Heidegger and Robinson Crusoe." In *The Ends of History: Questioning the Stakes of Historical Reason*, ed. Amy Swiffen and Joshua Nichols, 161–178. New York: Routledge.

———. 2012b. "To Die a Living Death: Phantasms of Burial and Cremation in Derrida's Final Seminar." *Societies* 2 (4): 317–331.

———. 2014. *The End of the World and Other Teachable Moments: Jacques Derrida's Final Seminar*. New York: Fordham University Press.

Nagel, Thomas. 1986. *The View from Nowhere*. Oxford: Oxford University Press.

Nancy, Jean-Luc. 1991. *Inoperative Community*. Minneapolis: University of Minnesota Press.

———. 1997. *The Sense of the World*. Minneapolis: University of Minnesota Press.

———. 2000. *Being Singular Plural*. Stanford: Stanford University Press.

———. 2007. *The Creation of the World, or, Globalization.* Albany: SUNY Press.

Napier, Susan J. 1993. "Panic Sites: The Japanese Imagination of Disaster from Godzilla to Akira." *Journal of Japanese Studies* 19 (2): 327–351.

Negrón, Mara. 2011. "Why Do Some Love Islands? Why Don't Others?" Comparative Caribbeans: An Interdisciplinary Conference, keynote address, November 5.

Nelson, Eric S. 2004. "Responding to Heaven and Earth: Daoism, Heidegger and Ecology." *Environmental Philosophy* 1 (2): 65–74.

Newman, Kim. 2000. *Apocalypse Movies: End of the World Cinema.* New York: St. Martin Griffin.

Nichols, Michael. 2013. *Earth to Sky: Among Africa's Elephants, a Species in Crisis.* New York: Aperture.

Nicks, Oran W., ed. 1970. *This Island Earth (NASA SP-250).* Washington, DC: Government Printing Office.

Nintzel, Jim. 2011. "Greenhouse Gift: The UA Takes Ownership of Biosphere 2—and Gets a $20 Million Grant to Help Cover Costs." *Tucson Weekly,* June 30. www.tucsonweekly.com/tucson/greenhouse-gift/Content?oid=3046843. Last Accessed July 12, 2013.

Norris, Christopher. 1986. "Home Thoughts from Abroad: Derrida, Austin, and the Oxford Connection." *Philosophy and Literature* 10 (1): 1–25.

Norris, Trevor. 2005. "Consuming Signs, Consuming the Polis: Hannah Arendt and Jean Baudrillard on Consumer Society and the Eclipse of the Real." *International Journal of Baudrillard Studies* 2 (2). www.ubishops.ca/baudrillardstudies/vol2_2/norrispf.htm.

———. 2006. "Hannah Arendt and Jean Baudrillard: Pedagogy in the Consumer Society." *Studies in Philosophy and Education* 25:457–477.

Nursoo, Ida. 2007. "Dialogue Across *Différance*: Hospitality Between Kant and Derrida." *Borderlands* 6 (3). www.borderlands.net.au/vol6no3_2007/nursoo_dialogue.htm.

Nwodo, Christopher S. 1976. "The Work of Art in Heidegger: A World Disclosure." *Philosophy and Social Criticism* 4 (1): 61–73.

O'Byrne, Anne. 2010. *Natality and Finitude.* Bloomington: Indiana University Press.

———. 2013. "*Communitas* and the Problem of Women." In *Angelaki* 18 (3): 125–138.

O'Connor, Patrick. 2007. "Derrida's Worldly Responsibility: The Opening Between 'Faith' and the 'Sacred.'" *Southern Journal of Philosophy* 45 (2): 303–334.

O'Connell-Rodwell, Caitlin. 2011. "Ritualized Bonding in Male Elephants." *New York Times,* July 20.

Odysseos, Louiza. 2009. "Constituting Community: Heidegger, Mimesis, and Critical Belonging." *Critical Review of International Social and Political Philosophy* 12 (1): 37–61.

Oliver, Kelly. 2001. *Witnessing: Beyond Recognition.* Minneapolis: University of Minnesota Press.

———. 2009. *Animal Lessons: How They Teach Us to Be Human.* New York: Columbia University Press.

———. 2011. "The Uncanny Strangeness of Maternal Election: Levinas and Kristeva on Parental Passion." In *Phenomenologies of the Stranger: Between Hostility and Hospitality*, ed. R. Kearney and K. Semonovotich. New York: Fordham University Press.

———. 2012. *Knock Me Up, Knock Me Down: Images of Pregnancy in Hollywood Films*. New York: Columbia University Press.

———. 2013. *Technologies of Life and Death: From Cloning to Capital Punishment*. New York: Fordham University Press.

Olkowski, Dorothea. 1999. "The Earth Screams; Life Itself." In *Gilles Deleuze and the Ruin of Representation*, 99–104. Berkeley: University of California Press.

Ortega, Mariana. 2001. "New Mestizas," "'World'-Travelers," and "Dasein": Phenomenology and the Multi-Voiced, Multi-Cultural Self. *Hypatia* 16 (3): 1–29.

———. 2005. "When Conscience Calls, Will Dasein Answer? Heideggerian Authenticity and the Possibility of Ethical Life." *International Journal of Philosophical Studies* 13 (1): 15–34.

Ott, Paul. 2009. "World and Earth: Hannah Arendt and the Human Relationship to Nature." *Ethics, Place, and Environment: A Journal of Philosophy and Geography* 12 (1): 1–16.

Overton, Rebecca, and Darrell Doods. 2006. "Lonely Only." *Horse and Rider* 45 (3): 52–73.

Parrett, Aaron. 2004. *The Translunar Narrative in the Western Tradition*. Burlington, VT: Ashgate.

Owens, Patricia. 2007. *Between War and Politics: International Relations and the Thought of Hannah Arendt*. Oxford: Oxford University Press.

Pasanen, Outi. 2006. "Notes on the *Augenblick* in and Around Jacques Derrida's Reading of Paul Celan's 'The Meridian.'" *Research in Phenomenology* 26 (1): 215–237.

Passerin d'Entreves, Maurizio. 1991. "Modernity and the Human Condition: Hannah Arendt's Conception of Modernity." *Thesis Eleven* 30:75–116.

Patterson, Bruce D. 2004. *The Lions of Tsavo: Exploring the Legacy of Africa's Notorious Man-Eaters*. New York: McGraw-Hill.

Patterson, John Henry. 1907. *The Man-Eaters of Tsavo*. London: Macmillan.

Pellegrini, Anthony D., Danielle Dupuis, and Peter K. Smith. 2007. "Play in Evolution and Development." *Developmental Review* 27 (2): 261–276.

Peterson, Anna. 2000. "In and of the World? Christian Theological Anthropology and Environmental Ethics." *Journal of Agricultural and Environmental Ethics* 12 (3): 237–261.

Phillips, Sam C. 1969. "Apollo 8: A Most Fantastic Voyage." *National Geographic* 135 (5): 593–635.

Platt, David. 1985. "The Seashore as Dwelling in the Fourfold." *International Philosophical Quarterly* 25 (2): 173–184.

Poole, Robert. 2008. *Earthrise: How Man First Saw the Earth*. New Haven: Yale University Press.

Potter, Claire. 2011. "Fishing and Thinking, or an Interiority of My Own." In *Thinking with Irigaray*, ed. Mary C. Rawlinson, Sabrina L. Hom, and Serene J. Khader, 201–220. Albany: SUNY Press.
Power, Marcus and Andrea Crampton. 2005. "Reel Geopolitics: Cinemato-graphing Political Space. *Geopolitics* 10 (2): 193–203.
Radloff, Bernhard. 2007. "Self-Overpowering Power and the Refusal of Being." *Existentia* 17 (5–6): 393–422.
———. 2012. "Crossing-Over and Going-Under: A Reading of Heidegger's Rectorial Address in Light of *Contributions to Philosophy*." *Heidegger Studies* 28:23–46.
Raffoul, Francois. 2008. "Derrida and the Ethics of the Im-possible." *Research in Phenomenology* 38 (2): 270–290.
Reinhardt, Olaf. 1982. "Kant's Thoughts on the Ageing of the Earth." *Annals of Science* 39 (4): 349–369.
Reinhardt, Olaf and D. R. Oldroyd. 1983. "Kant's Theory of Earthquakes and Volcanic Action." *Annals of Science* 40 (3): 247–272.
Relph, Edward. 1986. "Geographical Experiences and Being-in-the-World: The Phenomenological Origins of Geography." In *Dwelling, Place, and Environment*, ed. David Seamon and Robert Mugerauer, 15–31. Hingham, MA: Springer Netherlands.
Reshaur, Ken. 1992. "Concepts of Solidarity in the Political Theory of Hannah Arendt." *Canadian Journal of Political Science/Revue Canadienne de Science Politique* 25 (4): 723–736.
Rifkin, Jeremy. 2009. *The Empathic Civilization: The Race to Global Consciousness in a World in Crisis*. New York: Tarcher.
———. 2010. "'The Empathic Civilization': Rethinking Human Nature in the Biosphere Era." *Huffington Post*. www.huffingtonpost.com/jeremy-rifkin/the-empathic-civilization_b_416589.html.
Rigby, Kate. 2004. "Earth, World, Text: On the (Im)possibility of Ecopoiesis." *New Literary History* 35 (3): 427–442.
Rivlin-Nadler, Max. 2013. "More Than 78,000 People Have Applied to Die on Mars." *Gawker*, May 11. http://gawker.com/more-than-78-000-have-applied-to-die-on-mars-in-front-o-501686962. Last accessed July 12, 2013.
Robinson, Keith. 2010. "Towards a Political Ontology of the Fold: Deleuze, Heidegger, Whitehead, and the 'Fourfold' Event." In *Deleuze and the Fold: A Critical Reader*, ed. Sjoerd van Tuinen and Niamh McDonnell. New York: Palgrave Macmillan.
Rojcewicz, Richard. 1997. "Platonic Love: Dasein's Urge Toward Being." *Research in Phenomenology* 27 (1): 103–120.
Roney, Patrick. 2009. "Evil and the Experience of Freedom: Nancy on Schelling and Heidegger." *Research in Phenomenology* 39 (3): 374–400.
Rorty, Richard. 1994. "Another Possible World." In *Martin Heidegger: Politics, Art, and Technology*, ed. Karsten Harries and Christoph Jamme, 34–40. New York: Holmes and Meier.
Rose, Gillian. 1992. *The Broken Middle: Out of Our Ancient Society*. Oxford: Blackwell.

Rosner, David J. 2006. "Anti-Modernism and Discourses of Melancholy." *E-rea* 4 (1). DOI: 10.4000/erea.596.

Rottenberg, Elizabeth. 2006. "The Resistance to Interpretation." *Philosophy Today* 50:83–90.

Rousseau, Jean-Jacques. 1979. "Book 1." In *Emile*, ed. Allen Bloom, 37–74. New York: Basic Books.

Rudd, Anthony. 2003. "Part III: Heidegger and the External World." In *Expressing the World: Skepticism, Wittgenstein, and Heidegger*, 159–240. Chicago: Open Court.

Sagan, Carl. 1994. "Chapter 1: You Are Here." In *Pale Blue Dot: A Vision of the Human Future in Space*, 1–8. New York: Ballantine.

Salem-Wiseman, Jonathan. 2003. "Heidegger's Dasein and the Liberal Conception of the Self." *Political Theory* 31 (4): 533–557.

Sallis, John. 1998. "Levinas and the Elemental." *Research in Phenomenology* 28 (1): 152–159.

Schalow, Frank. 2006. *The Incarnality of Being: The Earth, Animals, and the Body in Heidegger's Thought*. Albany: SUNY Press.

Scheibler, Ingrid. 2001. "Art as Festival in Heidegger and Gadamer." *International Journal of Philosophical Studies* 9 (2): 151–175.

Schroeder, Brian. 2004. "The Inoperative Earth." *Studies in Practical Philosophy* 4 (1):126–145.

———. 2012. "Reterritorializing Subjectivity." *Research in Phenomenology* 42 (2): 251–266.

Shangquan, Gao. 2000. "Economic Globalization: Trends, Risks, and Risk Prevention." Committee for Development Policy Background Paper No. 1. *Economic and Social Affairs*. New York. New York.

Sharp, John G. and Jane C. Sharp. 2007. "Beyond Shape and Gravity: Children's Ideas About the Earth in Space Reconsidered." *Research Papers in Education* 22 (3): 363–401.

Sheldrick, Daphne. 2012. *Love, Life, and Elephants: An African Love Story*. New York: Farrar, Straus and Giroux.

Shuster, Martin. 2012. "Language and Loneliness: Arendt, Cavell, and Modernity." *International Journal of Philosophical Studies* 20 (4): 473–497.

Siebert, Charles. 2006. "An Elephant Crackup?" *New York Times Magazine*, October 8, 42–72.

———. 2011. "Orphan Elephants." *National Geographic*, September. http://ngm.nationalgeographic.com/2011/09/orphan-elephants/siebert-text.

Siegal, Michael, Gavin Nobes, and Georgia Panagiotaki. 2011. "Children's Knowledge of the Earth." *Nature Geoscience* 4:130–132.

Silverman, Hugh J. 2008. "Excessive Responsibility and the Sense of the World (Merleau-Ponty and Nancy)." *Chiasmi International* 10:307–318.

Sinaikin, Philip. 2009. "Bored to Tears? Depression and Heidegger's Concept of Profound Boredom: A Postpsychiatry Contribution." In *Philosophical Perspectives on Technology and Psychiatry*, ed. James Phillips, 181–203. Oxford: Oxford University Press.

Sinclair, Mark. 2006. *Heidegger, Aristotle, and the Work of Art: Poeisis in Being.* New York: Palgrave Macmillan.

Singh, R. Raj. 1990. "Heidegger and the World in an Artwork." *Journal of Aesthetics and Art Criticism* 48 (3): 215–222.

Skees, Murray. 2009. "The Lex Permissiva and the Source of Natural Right in Kant's Metaphysics of Morals and Fichte's Foundations of Natural Right." *International Philosophical Quarterly* 49 (3): 375–398.

Slaby, Jan. 2010. "The Other Side of Existence: Heidegger on Boredom." In *Habitus in Habitat II: Other Sides of Cognition*, ed. Sabine Flach and Jan Soffner. Bern: Peter Lang.

Smith, Mick. 2011. "Risks, Responsibilities, and Side Effects: Arendt, Beck, and the Politics of Acting Into Nature." In *Against Ecological Sovereignty: Ethics, Biopolitics, and Saving the Natural World*, 135–158. Minneapolis: University of Minnesota Press.

Smith, Quentin. 1978. "Scheler's Critique of Husserl's Theory of the World of the Natural Standpoint." *Modern Schoolman* 55 (4): 387–396.

Sontag, Susan. 1966. "The Imagination of Disaster." In *Against Interpretation*, 209–225. New York: Farrar, Straus and Giroux.

Stafford, Sue P. and Wanda Torres Gregory. 2006. "Heidegger's Phenomenology of Boredom and the Scientific Investigation of Conscious Experience." *Phenomenology and the Cognitive Sciences* 5 (2): 155–169.

Steger, Manfred B. 2003. *Globalization : A Very Short Introduction.* New York: Oxford University Press.

Stierle, Karlheinz. 1994. "An Eye Too Few: Earth and World in Heidegger, Holderlin, and Rousseau." In *Martin Heidegger: Politics, Art, and Technology*, ed. Karsten Harries and Christoph Jamme, 154–163. New York: Holmes and Meier.

Strohmayer, Ulf. 1998. "The Event of Space: Geographic Allusions in the Phenomenological Tradition." *Environment and Planning D: Society and Space* 16 (1): 105–121.

Stulberg, Robert B. 1973. "Heidegger and the Origin of the Work of Art: An Explication." *Journal of Aesthetics and Art Criticism* 32 (2): 257–265.

Sturm, Thomas. 2011. "Freedom and the Human Sciences: Hume's Science of Man Versus Kant's Pragmatic Anthropology." *Kant Yearbook* 3:23–42.

Sundararajan, Louise. 1999. "Being as Refusal: Melville's Bartleby as Heideggerian Anti-Hero." *Janus Head*. www.janushead.org/jhsumm99/sundararajan.cfm. Last Accessed March 12, 2013.

Szerszynski, Bronislaw. 2003. "Technology, Performance, and Life Itself: Hannah Arendt and the Fate of Nature." *Sociological Review* 51 (s2): 203–218.

Thiele, Leslie Paul. 1995. *Timely Meditations: Martin Heidegger and Postmodern Politics.* Princeton: Princeton University Press.

———. 1997. "Postmodernity and the Routinization of Novelty: Heidegger on Boredom and Technology." *Polity* 29 (4): 489–517.

Thompson, Kirsten Moana. 2007. *Apocalyptic Dread: American Film at the Turn of the Millennium.* Albany: SUNY Press.

Thomson, Iain. 1999. "Can I Die? Derrida on Heidegger on Death." *Philosophy Today* 43 (1): 29–42.

———. 2011. "Heidegger's Postmodern Understanding of Art." In *Heidegger, Art, and Postmodernity*, 65–120. Cambridge: Cambridge University Press.

Tijmes, Pieter. 1998. "Home and Homelessness: Heidegger and Levinas on Dwelling." *Worldviews: Global Religions, Culture, and Ecology* 2 (3): 201–213.

Time. 1969. "Men of the Year." *Time*, January 3, 1.

Toadvine, Ted. 2005. "Limits of the Flesh: The Role of Reflection in David Abram's Ecophenomenology." *Environmental Ethics* 27 (2): 155–170.

Turnbull, Neil. 2006. "The Ontological Consequences of Copernicus: Global Being in the Planetary World." *Theory Culture Society* 23:125.

"Two Former Biosphere Workers Are Accused of Sabotaging Dome." 1994. *New York Times*, April 5. www.nytimes.com/1994/04/05/us/two-former-biosphere-workers-are-accused-of-sabotaging-dome.html. Last Accessed July 12, 2013.

Vallega-Neu, Daniela. 2003. *Heidegger's Contributions to Philosophy: An Introduction*. Bloomington: Indiana University Press.

Vallier, Robert. 2009. "Elemental Difference: Of Life, Flesh, and Earth in Merleau-Ponty and the Timaeus." In *Merleau-Ponty and the Possibilities of Philosophy: Transforming the Tradition*, ed. Robert Vallier, Wayne Jeffrey Froman, and Bernard Flynn, 129–154. Albany: SUNY Press.

Vatter, Miguel. 2006. "Natality and Biopolitics in Hannah Arendt." *Revista de Cienca Politica* 26 (2): 137–159.

Vedder, Ben. 1998. "Heidegger on Desire." *Continental Philosophy Review* 31:352–368.

———. 2005. "A Philosophical Understanding of Heidegger's Notion of the Holy." *Epoché* 10 (1): 141–154.

Vernadsky, Vladimir I. 1998. *The Biosphere*. Trans. D. B. Langmuir. New York: Copernicus.

Vygotsky, Lev S. 1978. "The Role of Play in Development." In *Mind in Society*, trans. M. Cole, 92–104. Cambridge: Harvard University Press.

Waal, Frans de. 2006. *Our Inner Ape: A Leading Primatologist Explains Why We Are Who We Are*. New York: Riverhead.

———. 2009. *The Age of Empathy: Nature's Lessons for a Kinder Society*. New York: Broadway.

Walliss, John and James Aston. 2011. "Doomsday America: The Pessimistic Turn of Post-9/11 Apocalyptic Cinema." *Journal of Religion and Popular Culture* 23 (1): 53–64.

Waterworth, Tanya. 2012. "Elephants Say Goodbye to the Whisperer." *IOL News*, March 10. www.iol.co.za/news/south-africa/kwazulu-natal/elephants-say-goodbye-to-the-whisperer-1.1253463#.U7rMzLFCy_Q.

Weber, Elisabeth. 2005. "Deconstruction Is Justice." *SubStance* 34, no. 1, 106:38–43.

Weber, Ronald. 1985. *Seeing Earth: Literary Responses to Space Exploration*. Athens: Ohio University Press.

Weber, Samuel. 1989. "In the Name of the Law." *Cardozo Law Review* 11:1515–1538.

Weinberger, David. 1984. "Earth, World and Fourfold in the Thought of Martin Heidegger." *Tulane Studies in Philosophy* 32:103–109.
Westmoreland, Mark W. 2008. "Interruptions: Derrida and Hospitality." *Kritike* 2 (1): 1–10.
Westphal, Kenneth R. 1997. "Do Kant's Principles Justify Property or Usufruct?" *Annual Review of Law and Ethics* 5:141–194.
———. 2002. "A Kantian Justification of Possession." In *Kant's Metaphysics of Ethics: Interpretive Essays*, ed. M. Timmons, 89–109. Oxford: Oxford University Press.
Whiteside, Kerry. 1994. "Hannah Arendt and Ecological Politics." *Environmental Ethics* 16 (4): 339–358.
Wigley, Mark. 1987. "Postmortem Architecture: The Taste of Derrida." *Perspecta* 23: 156–172.
Wilkerson, Dale Allen. 2005. "The Root of Heidegger's Concern for the Earth at the Consummation of Metaphysics: The Nietzsche Lectures." *Cosmos and History: The Journal of Natural and Social Philosophy* 1 (1): 27.
Willett, Cynthia. 1995. *Maternal Ethics and Other Slave Moralities*. New York: Routledge.
———. 2001. *The Soul of Justice: Social Bonds and Racial Hubris*. Ithaca: Cornell University Press.
———. 2008. *Irony in the Age of Empire*. Bloomington: Indiana University Press.
———. 2014. *Interspecies Ethics*. New York: Columbia University Press.
Wilson, Edward O. 1993. "Biophilia and the Conservation Ethic." In *The Biophilia Hypothesis*, ed. Stephen R. Kellert and Edward O. Wilson, 31–41. Washington, DC: Island.
Wilson, Holly L. 2006. *Kant's Pragmatic Anthropology: Its Origin, Meaning, and Critical Significance*. Albany: SUNY Press.
———. 2011. "The Pragmatic Use of Kant's Physical Geography Lectures." In *Reading Kant's Geography*, ed. Stuart Elden and Eduardo Mendieta, 161–172. Albany: SUNY Press.
Winerip, Michael. 2013. "Good Science, or Bad Sense?" *New York Times*, June 10. www.nytimes.com/2013/06/10/booming/biosphere-2-good-science-or-bad-sense.html?hp&_r=0. Last accessed July 1, 2013.
Winkler, Rafael. 2007. "Heidegger and the Question of Man's Poverty in World." *International Journal of Philosophical Studies* 15 (4): 521–539.
Winning, Anne. 1990. "Homesickness." *Phenomenology and Pedagogy* 8:245–258.
Winter, Christine. 2010. "Places, Spaces, Holes for Knowing and Writing the Earth: The Geography Curriculum and Derrida's Khôra." *Ethics and Education* 4 (1): 57–68.
Withy, Katherine. 2013. "The Strategic Unity of Heidegger's *The Fundamental Concepts of Metaphysics*." *Southern Journal of Philosophy* 51 (2): 161–178.
Wolfe, Cary. 2009. "Exposures." In *Philosophy and Animal Life*, 1–42. New York: Columbia University Press.
Wolin, Richard. 2003. *Heidegger's Children: Hannah Arendt, Karl Lowith, Hans Jonas, and Herbert Marcuse*. Princeton: Princeton University Press.
———. 2010. "The Idea of Cosmopolitanism: From Kant to the Iraq War and Beyond." *Ethics and Global Politics* 3 (2): 143–153.

Wood, David. 1999. "*Comment ne pas manger*—Deconstruction and Humanism." In *Animal Others: On Ethics, Ontology and Animal Life*, ed. Peter Steeves, 15–35. Albany: SUNY.
———. 2002. *Thinking After Heidegger*. Malden, MA: Polity.
———. 2004. "Thinking with Cats." In *Animal Philosophy: Ethics and Identity*, ed. Peter Atterton and Matthew Calarco. New York: Continuum.
———. 2006. "On the Way to Econstruction." *Environmental Philosophy* 3 (1): 35–46.
———. 2012. "The Truth About Animals." *Environmental Philosophy* 9 (2): 159–167.
Yates, Joshua J. 2009. "Mapping the Good World: The New Cosmopolitans and Our Changing World Picture." *Hedgehog Review* 11 (3): 7–27.
Young, Julian. 2000. "What Is Dwelling? The Homelessness of Modernity and the Worlding of the World." In *Heidegger, Authenticity, and Modernity: Essays in Honor of Hubert L. Dreyfus*, ed. Mark Wrathall and Jeff Malpas, 1:187–204. Cambridge: MIT Press.
———. 2001. *Heidegger and Modern Art*. Cambridge: Cambridge University Press.
———. 2004. *Heidegger's Philosophy of Art*. Cambridge: Cambridge University Press.
———. 2005. "Death and Transfiguration: Kant, Schopenhauer, and Heidegger on the Sublime." *Inquiry* 48 (2): 131–144.
———. 2006. "The Fourfold." In *The Cambridge Companion to Heidegger*, ed. Charles Guignon, 373–392. 2d ed. Cambridge: Cambridge University Press.
Young-Bruehl, Elisabeth. 2004. *Hannah Arendt: For Love of the World*. New Haven: Yale University Press.
Yu, Xuanmeng. 2007. "The Homesickness or Nostalgia of a Modern People." In *Shanghai: Its Urbanization and Culture*, ed. Xuanmeng Yu and Xirong He. Washington, DC: CRVP.
Zammito, John H. 2008. "A Text of Two Titles: Kant's 'A Renewed Attempt to Answer the Question: "Is the Human Race Continually Improving?"'" *Studies in History and Philosophy of Science* 39:535–545.
Zerilli, Linda M. G. 2005. "'We Feel Our Freedom': Imagination and Judgment in the Thought of Hannah Arendt." *Political Theory* 33:158.
Ziarek, Krzysztof. 2000. "Proximities: Irigaray and Heidegger on Difference." *Continental Philosophy Review* 33 (2): 133–158.
Zimmerman, Michael E. 1983. "Toward a Heideggerean Ethos for Radical Environmentalism." *Environmental Ethics* 5 (2): 99–131.
———. 1993. "Rethinking the Heidegger—Deep Ecology Relationship." *Environmental Ethics* 15 (Fall): 195–224.
———. 1994. *Contesting Earth's Future: Radical Ecology and Postmodernity*. Berkeley: University of California Press.
Zlomislić, Marko. 2007. *Jacques Derrida's Aporetic Ethics*. Lanham, MD: Lexington.
Zuidervaart, Lambert. 2002. "Art, Truth, and Vocation: Validity and Disclosure in Heidegger's Anti-aesthetics." *Philosophy and Social Criticism* 28 (2): 153–172.

INDEX

acquisition, 50, 52–60
action, 75—83
Allen, Wayne, 252n16
All Is Lost, 207–11, 237, 244
amor mundi, 32, 72, 103, 104, 241
amor terra, 32, 241
Anders, William, 3, 13, 16, 248n21
animal laborans, 74, 75
animals, 34, 74, 78, 104, 118, 121, 141, 173, 217, 232; access to death, 176, 177, 204; as historical, 78; as a pedagogical tool, 171; as political, 80; poor in world, 117, 122; resources for labor, 213; sharing human world, 74; uncanny, 121; world-building, 170; *see also* earthlings
annihilation: anniliation of worldview, 84, 88; earth annihilation, 10–12, 157, 239; total annihilation, 3, 32, 34, 107; world annihilation, 73, 93
Anthony, Lawrence, 204
Apollo missions, 2, 3, 6, 10, 12, 15–17, 159, 164, 208, 211, 228, 239, 243
Arendt, Hannah, 4–11, 25–31, 37, 41, 43, 71–110, 114, 124, 138, 144, 165–69, 189, 194, 199, 212–19, 224, 228, 239, 241

art, 32, 33, 99, 123, 126, 128, 159, 191, 200, 255
attunement, 116–17, 120, 218
Augustine, Aurelius 104–6, 225,
authenticity, inauthentic relation to the world, 115, 116
autoimmune logic, 12, 19, 22, 29, 95, 191, 215, 216; island, 164; Robinson Crusoe, 36, 183; sovereignty, 183; terrorism, 36; wheel, 182

Bender, Hans, 202
Benhabib, Seyla, 262n29
Berger, John, 119
Bernasconi, Robert, 255n19, 259n10,
biodiversity, 4, 9, 25, 33, 41–42, 72, 91, 102, 140, 232, 240, 241; implications for politics, 106; *see also* ecosystem
bios, 75, 81
biosphere, 27, 39, 40, 158, 161, 212, 219–20, 227-44 *passim*
Biosphere 2, 27, 28
Birmingham, Peg, 252n18
Blue Marble, 1, 11, 14, 19, 20, 25, 224; *see also* Apollo Missions

Bond, James: *The Day the Earth Caught Fire*, 10; *You Only Live Twice*, 10
boredom, 120, 175
Borman, Frank, 14, 16
Bowlby, John, 262n21
Buchanan, Brett, 254n5
Butler, Judith, 83–84, 252n15

Caputo, John, 254n8
care, 112, 116
Celan, Paul, 11, 171, 173, 174, 175, 187, 188, 199, 259n9, 259n12–13, 260n32
Cerbone, David, 254n3
Chandor, C. J., (*All Is Lost*), 207
Chiba, Shin, 253n28, 253n32, 253n33
Clark, David, 251n27
climate change, 21, 78, 107, 150–53, 160, 226, 237, 240
cohabitation, 102, 109, 113, 138, 220–30; choosing the unchosen, 83–84, 225; earth ethics, 139; human condition, 212, 218; interspecies cohabitation, 233; nonhuman animals, 85, 168, 213; sharing the earth, 85; sharing the world, 89; with Eichmann, 84; *see also* plurality
Cohen, Alix, 250n3
Cohen, Richard, 259n10
cold war, 2, 6, 10, 14, 15, 25, 109, 159, 165, 237
Colebrook, Claire, 248n31
Collins, Michael 15, 16
common possession, 5, 7, 42, 46, 99, 168; earth's surface, 53, 55, 68, 69, 218
common world, 89, 91, 180, 181, 185; Eichmann, 199; language, 190, 192, 199; prosthesis, 182, 184, 203, 217, 259; stabilizing apparatuses, 190, 180
concealment, 122–26, 133, 134, 135, 142, 151, 154, 158
Corman, Roger (*The Day The World Ended*), 10
Cosgrove, Dennis, 22, 248n16
cosmopolitanism, 5, 6, 7, 14, 46, 72, 106, 226, 240; based on singularity, 169; human solidarity, 106; nationalism, 7, 165, 217;

perpetual peace, 46, 51; tension between conditional and unconditional, 165; tyranny, 77
Cronkite, Walter, 248n22
Crusoe, Robinson, 9, 18, 35, 37, 163, 164, 166, 169, 175, 179, 183, 207
Cuarón, Alfonso, 245

Dasein, 112, 115, 121, 174, 214; fourfold, 140; home, 113, 136, 115; protector of earth, 130, 141; thrownness, 113, 116; world-building, 175; world-forming, 116, 117
Darwin, Charles, 212, 230, 231
death, 142, 171, 172; animal, 176, 177, 204; as such, 177; belonging to others, 178; definitive of world, 178; deprived of world, 194; end of the world, 187, 193; fetish, 188; mortality, 173, 174; mourning, 178; of the other, 175; pleasure, 178; stabilizing structures, 194; world and war, 188
Derrida, Jacques, 4–11, 18, 19, 22, 28–31, 36, 37, 42, 66, 87, 95, 106, 109, 119, 133, 163–206, 208, 212, 215–18, 221–23, 235, 239, 241, 259n9, 259n15, 259n24, 260n26
desert, 37, 103, 106; desertification, 33, 34; oasis, 105; totalitarianism, 32
de Waal, Frans, 230, 231
Dietz, Mary, 252n9
dissembling, *see* refusal
Donne, John, 178, 182
dwelling, 116, 125, 137; as earthlings, 160; building, 143; earth's refusal, 157; fourfold, 142; guarding, 138; habitation, 139; loving earth, 144; poetic, 158, 175; possibility for response, 148; questioning, 157; restraint, 158; safeguarding, 149

earth ethics, 5, 40, 42, 43, 72, 106, 123, 203, 206, 216, 218, 221–22, 243; adopting earth as home, 138, 195; alternative to technological enframing, 114; based on cohabitation, 84, 41–42, 104, 213; based on limitation, 112, 215; earthlings, 5,

8, 31, 39, 219; ethics of conservation, 106, 110; hermeneutic ethics, 139, 157; hospitality, 169; hyperbolic ethics, 221; relationality, 39, 222; responsibility, 226, 235–36, 240; singular bond, 223, 234; sustainable ethics, 110; *terraphilia*, 208, 212, 224, 228, 232

earthlings, 28, 39, 48, 95, 102, 104–5, 113, 208, 213, 218; belonging to earth, 28–29, 225; cohabitation, 48, 160; singular bond, 84, 240

Earthrise, 1, 3, 13, 15, 20, 239, 244, 246

ecosystem, 7, 9, 36, 42, 58, , 97, 221, 222, 226, 235; *see also* biodiversity

Edwards, Jeffrey, 53, 250n8, 250n9, 251n13

ego, 185, 186, 203, 217, 229, 230

Eichmann, Adolf, 8, 11, 83–85, 93, 189, 195, 199, 226

environmental crisis, 78, 106, 107, 109; environmental management, 25; saving power, 160, 239, 240

equilibrium, 10, 50, 65–66, 166, 221, 230

Felman, Shoshana, 94, 199, 200
Fenves Peter, 249n2
The Final War, 10
Foltz, Bruce, 256n35, 256n40, 258n71
Foucault, Michel, 179
fourfold, 112, 118, 125, 140, 142; concealing and revealing, 113, 146, 151; divinities, 145, 146, 147; limit of responsibility, 144
Freud, Sigmund, 179, 229, 230, 254n6
Fried, Gregory, 255n20–21, 255n26, 256n33
Fuller, Buckminster 15

Gadamer, Hans-Georg, 127, 152, 193, 200, 255n27, 255n30
givenness, 8, 81, 95, 102, 113, 125, 128; pregiveness, 21, 22, 126, 127, 147
globalization, 4, 25, 30, 114, 135, 208; consumerism, 25; fantasy of *whole earth*, 11; global telecommunication, 4, 15, 153; globe, 12; planetary imperialism, 31; planetary thinking, 25, 155; technoscience, 14; totalitarianism, 28; uprootedness, 152; world citizenship, 5; worldlessness, 31

Gopnik, Alison, 240
Gravity, 207, 211, 237, 244–46

Haar, Michel, 255n19
habitation, 7, 47, 56, 57, 114, 138–40, 167, 199; earth, 87; world, 85, 220
Harper, Jack (Tom Cruise), 34, 35, 237, 238, 246
Harris, John 99, 127
Heidegger, Martin, 4–5, 7–9, 11, 17, 22, 25, 27–41, 106, 108, 111–61, 164, 168–72, 174–77, 184, 189, 194, 196, 200, 202, 212–19, 224, 233, 236, 240, 256n32; *Contributions*, 256n41; death, 248n7–8; *Fundamental Concepts of Metaphysics*, 259n9
Hedrick, Todd, 250n6
Held, Klaus, 254n3
history, 63, 64, 134, 138, 250, 263, 264, 265, 268, 271, 273, 274, 276, 277, 278, 280, 281, 283, 284, 287, 291
Hölderlin, Friedrich, 156, 254n11, 260n34
Hobbes, Thomas, 39
home, 9, 78–79, 90, 113, 118, 124, 160, 234, 242, 262n20; ambiguity, 136, 138, 168; animals, 76; dwelling, 109, 112; earth, 78, 112, 219; environmental crisis, 78; habitation, 114; homelessness, 78, 100, 119, 157, homesickness, 117, 120, 176, 179; human artifice, 74–75; love, 112, 234; philosophy's desire, 118; private property, 76; stateless people, 98; uncanny, 168, 238
Honig, Bonnie, 253n31, 258n3
hospitality, 6, 7, 165, 167, 184, 212, 216, 226; asocial sociability, 166, island, 166; inhospitable hospitality, 6, 47, 61, 166; cosmopolitan right, 61; limited surface of the earth, 6, 7, 46; perpetual peace, 7, 11, 45; poetic affirmation, 169; right of visitation, 6, 45; unconditional hospitality, 9, 166
Homer, 88, 91
homo faber, 75, 76; human artifice, 73
human condition, 6, 26–27, 76, 82

human rights, 7, 28, 73, 75–78, 82, 90–91, 95, 98–99, 101–4, 213, 218, 225
human species, 13, 52, 53, 58, 62–64, 69, 72, 213, 227, 230; human solidarity, 107–9
Husserl, Edmund, 20, 21, 26, 28, 89, 90, 116, 127, 218, 219, 233, 248n28

imagination, 26, 72, 90, 101, 102, 109, 110, 226
international law, 7, 28, 72, 107
interpretation, 23, 24, 56, 57, 58, 86, 88, 102, 113, 123, 125–28, 133–40, 154, 202, 203, 206, 225, 235; openness to beings, 137; poetic 192, 196, 199, 217; saving power, 149, 156; shifting perspective, 148
ipseity, 182–85; ego, 186; illusion of autonomy, 185
Iraq, 11, 187
island, 10, 17, 18, 19, 163, 166, 169, 173, 179, 215; isolated and unique, 164; language, 192; metaphor, 39; singularity, 185, 223; testimony, 186

Jaspers, Karl, 27, 106–8
Jefferson, Thomas, 99
Joronen, Mikko, 257n64

Kant, Immanuel, 4–11, 25, 28, 31, 32, 38, 41–43, 45–69, 71, 73, 81, 92, 100, 106, 107, 109, 111–14, 125, 138, 165, 168, 196, 208, 212, 215–27, 234, 247n12, 250n4, 250n7, 251n12, 251n16
Kateb, George, 253n26
Kierkegaard, Søren, 216
King, Jr., Martin Luther, 225
Kleingeld, Pauline, 250n6
Kohut, Heinz, 262n21
Kosinski, Jospeh, 236
Kowalski, Matt (George Clooney), 244–46
Kramer, Stanley (*On the Beach*), 10, 188
Krell, David, 143, 258n4
Kristeva, Julia, 18, 103, 192
Kubrick, Stanley (*Dr. Strangelove*), 10

labor, 75, 96–98, 213
Lacan, Jacques, 24

Lacoue-Labarthe, Philippe, 257n58
Lazier, Benajmin, 12, 153, 241n1, 248n34
Laub, Dori, 93, 94
Lévinas, Emmanuel, 105, 197, 200, 201
limit, 73, 95, 100, 144–45; between social and political, 97 earth alienation, 96; earth as a limit concept, 75, 103, 111, 212; earth as limit of world, 214, 219; fourfold, 140, 142, 213; genocide, 88; human perspective, 239; limited resources, 218; private and public realms, 80; surface of the earth, 53, 212, 220, 242; war, 81, 82; world alienation, 96, 97
Locke, John, 39
loneliness, 12, 15–16, 38, 119, 122, 164, 238
Louden, Robert, 250n6
love, 32, 33, 43, 72, 103–6, 109, 144, 163, 169, 224–26, 229, 232; *Agape*, 229; *biophilia*, 227, 232; creativity, 229; empathy, 231; *eros*, 229, 230, 233, 234, 236; *terraphilia*, 208, 228
Lovell, James, 13, 16, 18, 38
Lovelock, James, 227
Low, George, 248n23
Lunar Orbiter I, 3, 33, 152, 159

Macauley, David, 254n23
MacLean, Paul, 230
MacLeish, Archibald, 3, 13, 17, 165
Madison, James, 99
Mann, Bonnie, 252n20
Mars, 18, 27, 28, 68, 100, 152, 153,
Mars One, 18, 27, 28, 100, 153
mastery, 16, 23, 24, 39–42, 111, 125, 131–34, 143, 147, 156, 229, 235, 240, 241, 246; divinities, 145; political sovereignty, 189, 191
McNeill, William, 254n11; *The Time of Life*, 261n8
Merleau-Ponty, Maurice, 262n19
Miller, Elaine, 64
Mitchell, Andrew, 33, 34, 149, 249n43–47, 255n13, 255n17, 255n28, 256n45–46, 256n49–50, 257n53, 257n62, 258n2, 262n22

mondialisation, *see* globalization
moral law, 48, 49

Naas, Michael, 185, 186, 258n4, 259n16–19
Nancy, Jean-Luc, 257n65, 261n7
narratives, 75, 89, 92, 93, 95; genocide, 92
NASA, 1, 4, 13–16, 19, 166, 243
natality, 73, 77, 79–80, 86, 103, 106, 186, 219, 229; action, 79; creativity, 105; nature and history, 86
national rights, 6
National Socialism, 11, 73, 83, 85, 88, 94–96, 127
nation-states, 6, 10, 11, 29, 46–48, 51, 64, 72, 76–78, 107, 168
natural law, 49, 99
Novalis, 117, 171, 172, 174–76, 178; *see also* philosophy
nuclear war, 3, 10–12, 24, 29, 81, 109, 150, 238

O'Byrne, Anne, 86, 90, 92, 249n50, 251n28
Oblivion, 12, 34, 207, 236–38, 246
One-World, 14, 15, 23, 28, 159
Ott, Paul, 253n21, 253n23

Paine, Thomas, 13
Pathfinder, 68
Patterson, John Henry (Colonel Patterson), 57
perfectability, 47, 48, 53, 66, 196
perpetual peace, 28, 45; cosmological peace, 69; equilibrium of political forces, 50; perpetual war, 50; private property, 47
perspective, 89, 81, 101; alien perspective, 10, 68; Archimedean vantage point, 26; Einstein's free-floating observer, 98; God's eye view, 19, 20; human vantage point, 20, 22; other species, 38; whose perspective counts, 92
philosophy, 32, 117, 120, 171, 174–78, 196; homesickness, 120, 137, 171
planetary imperialism, 31, 40, 111, 112, 125, 136, 241
Plato, 105, 229, 260n30
pleasure, 178–79

plurality, 7, 25, 72, 84, 90, 102, 105, 108, 194, 212, 218, 220; *biophilia*, 233; cohabitation, 73; dynamic equilibrium, 221; law of the earth, 106; limit, 77; possibility of a world, 90; singularity, 80
poetry, 192, 195, 196, 199–202, 205, 217; poetic sovereignty: 218
political right, 5, 8, 28, 58, 76, 81; doctrine of right, 49
politics, 4–9, 32, 42, 47, 48, 50, 72, 75–77, 82–84, 95, 103, 106, 113, 167, 206, 208, 214–16, 223, 235
private property, 47, 50, 72, 78; original acquisition, 54, 55; provisional, 51, 52;
private realm, 80, 96
prostheses, *see* stablizing apparatuses
public realm, 80, 96

Radloff, Bernhard, 254n8
real world, 89, 90–92; *see also* common world
reason, 43, 46–49, 50, 52, 53, 62–69, 149, 212, 221, 234, 236
Redford, Robert, 207, 208
refusal, 125, 126, 132, 134, 156, 214, 223; animal refusal, 121; concealment, 122, 123; dwelling, 157; gift, 158; revealing, 123, 125; strife, 133; telling refusal, 130, 131
Reinhardt, Otto, 239n2
relationality, 8, 9, 33, 37, 105, 112, 113, 118, 123, 128, 131, 140, 141, 142, 146, 155, 214, 215, 216, 220, 222, 229, 232, 234, 235, 241; dependence, 143; embodied, 236; relational phenomenology, 140; singularity, 179; without hierarchy, 142
responsibility, 141, 143, 149, 172, 192, 202, 218, 223; earth, 208, 221, 228; environment, 157; hyperbolic ethics, 112, 198; poetry, 200; questioning, 151; singular bond with the earth, 220
restraint, 101, 112, 129–34, 143, 145, 158–59
Rifkin, Jeremy, 231, 261n10–13, 262n21
Rousseau, Jean-Jacques, 261n5

Sagan, Carl, 17, 18, 160, 243

Sallis, John, 261n3, 261n6
saving power, 40, 149, 150–52, 156, 158–60, 208, 239, 240; *see also* technological worldview
Schalow, Frank, 255n26
Schroeder, Brian, 261n7
self-seclusion, 132, 145, 214
Shuangquan, Gao, 249n40, 256n36
Siebert, Charles, 205
Smith, Mick, 258n71
Socrates, 130
sovereignty, 19, 35, 37, 73, 77, 133, 167, 169, 183, 184, 189, 190, 200, 202, 215, 216, 225, 229, 236; poetic sovereignty, 191, 206, 217; political sovereignty, 191, 206; recogntion, 225
space colonization, 4, 11, 18, 20, 21, 27, 99, 159
speech, 75, 83, 94
Der Spiegel, 144, 152
Sputnik, 2, 3, 9, 27, 28, 153
stabilizing apparatuses, 180, 181, 185, 190, 194, 196–98, 203, 204, 217
statelessness, *see* home
Stone, Ryan (Sandra Bullock), 244–46
strife, 116, 123, 126, 127, 128, 133; relation to home, 137; restraint, 129; tension between history and destiny, 125
Supreme Court: The Myriad Corporation, 99, 100
syncopated temporality, 86, 92

technological worldview, 11, 24, 28, 34, 111, 112, 124, 135, 144, 149, 155; access to weapons, 30; alienation, 100; complacency, 27; fantasy of wholeness, 25; homelessness, 137; planetary determinism, 153; standing reserve, 144, 150; technological enframing, 137, 151; technoscience, 29, 98, 133, 134, world domination, 129

terraphilia, 208, 212, 224, 228, 232; *amor terra*, 241
terrorism, 11, 29, 36, 150; Al-Qaeda, 10; World Trade Center, 11; *see* globalization
Thomson, Iain, 255n26, 259n10
totalitarianism, 28, 31, 77, 98, 107

uncanny, 9, 34–36, 39, 42, 105, 112, 118–22, 136,152, 169, 191, 200, 221, 222, 238, 239, 240
understanding heart, 101, 102, 109

van Gogh, Vincent, 256n32
Vedder, Ben, 257n55
Vernadsky, Vladimir, 60, 251n20
vulnerability, 15, 20

Walker, Nicholas, 254n11
war, 82, 94; just war, 72, 82, 87; total war, 72, 73, 82, 83, 87, 93, 94, 107, 108; world war, 187, 189
Westphal, Kenneth, 250n10
Whole Earth, 12, 15, 16, 20, 22, 28, 159
Willett, Cynthia, 192, 233, 235, 262n16–18, 262n23–24; *Interspecies Ethics*, 262n21; *Maternal Ethics*, 262n21
Wilson, Edward O., 227, 232
Wilson, Holly, 250n6
Winnicott, Donald, 262n21
witnessing, 42, 72, 82, 92–95, 186, 192, 199, 218, 236
Wolin, Richard, 250n6
Wood, David, 255n31, 261n9, 262n27
work, 75, 96
worldlessness, 33, 81, 95, 97, 103, 117, 172, 175
worldliness, 104, 115

Young, Julian, 255n12, 256n49–50, 257n53

zoë, 75, 81

GPSR Authorized Representative: Easy Access System Europe, Mustamäe tee 50, 10621 Tallinn, Estonia, gpsr.requests@easproject.com

www.ingramcontent.com/pod-product-compliance
Lightning Source LLC
Chambersburg PA
CBHW021936290426
44108CB00012B/856